AF889983

HIGH LEVEL SYNTHESIS OF ASICs UNDER TIMING AND SYNCHRONIZATION CONSTRAINTS

THE KLUWER INTERNATIONAL SERIES IN ENGINEERING AND COMPUTER SCIENCE

VLSI, COMPUTER ARCHITECTURE AND DIGITAL SIGNAL PROCESSING

Consulting Editor
Jonathan Allen

Latest Titles

Hardware Annealing in Analog VLSI Neurocomputing, B. W. Lee, B. J. Sheu
ISBN: 0-7923-9132-2
Neural Networks and Speech Processing, D. P. Morgan, C.L. Scofield
ISBN: 0-7923-9144-6
Silicon-on-Insulator Technology: Materials to VLSI, J.P. Colinge
ISBN: 0-7923-9150-0
Microwave Semiconductor Devices, S. Yngvesson
ISBN: 0-7923-9156-X
A Survey of High-Level Synthesis Systems, R. A. Walker, R. Camposano
ISBN: 0-7923-9158-6
Symbolic Analysis for Automated Design of Analog Integrated Circuits, G. Gielen, W. Sansen,
ISBN: 0-7923-9161-6
High-Level VLSI Synthesis, R. Camposano, W. Wolf,
ISBN: 0-7923-9159-4
Integrating Functional and Temporal Domains in Logic Design: The False Path Problem and its Implications, P. C. McGeer, R. K. Brayton,
ISBN: 0-7923-9163-2
Neural Models and Algorithms for Digital Testing, S. T. Chakradhar, V. D. Agrawal, M. L. Bushnell,
ISBN: 0-7923-9165-9
Monte Carlo Device Simulation: Full Band and Beyond, Karl Hess, editor
ISBN: 0-7923-9172-1
The Design of Communicating Systems: A System Engineering Approach, C. J. Koomen
ISBN: 0-7923-9203-5
Parallel Algorithms and Architectures for DSP Applications, M. A. Bayoumi, editor
ISBN: 0-7923-9209-4
Digital Speech Processing: Speech Coding, Synthesis and Recognition A. Nejat Ince, editor
ISBN: 0-7923-9220-5
Sequential Logic Synthesis, P. Ashar, S. Devadas, A. R. Newton
ISBN: 0-7923-9187-X
Sequential Logic Testing and Verification, A. Ghosh, S. Devadas, A. R. Newton
ISBN: 0-7923-9188-8
Introduction to the Design of Transconductor-Capacitor Filters, J. E. Kardontchik
ISBN: 0-7923-9195-0
The Synthesis Approach to Digital System Design, P. Michel, U. Lauther, P. Duzy
ISBN: 0-7923-9199-3
Fault Covering Problems in Reconfigurable VLSI Systems, R. Libeskind-Hadas, N. Hasan, J. Cong, P. McKinley, C. L. Liu
ISBN: 0-7923-9231-0

HIGH LEVEL SYNTHESIS OF ASICs UNDER TIMING AND SYNCHRONIZATION CONSTRAINTS

by

David C. Ku
Redwood Design Automation
&
Stanford University

and

Giovanni De Micheli
Stanford University

Kluwer Academic Publishers
Boston/Dordrecht/London

Distributors for North America:
Kluwer Academic Publishers
101 Philip Drive
Assinippi Park
Norwell, Massachusetts 02061 USA

Distributors for all other countries:
Kluwer Academic Publishers Group
Distribution Centre
Post Office Box 322
3300 AH Dordrecht, THE NETHERLANDS

Library of Congress Cataloging-in-Publication Data

Ku, David C., 1964
High level synthesis of ASICs under timing and synchronization constraints / by David C. Ku and Giovanni De Micheli.
p. cm. -- (Kluwer international series in engineering and computer science ; 177)
Includes bibliographical references and index.
ISBN 0-7923-9244-2
1. Application specific integrated circuits--Design and construction--Data processing. 2. computer-aided design. I. De Micheli, Giovanni. II. Title. III. Series: Kluwer international series in engineering and computer ; SEC 177.
TK874.6.K8 1992
621.381'5--dc20 92-10571
CIP

Printed on acid-free paper.

Printed in the United States of America

To our loving families.

Contents

Preface

Computer-aided synthesis of digital circuits from behavioral level specifications offers an effective means to deal with the increasing complexity of digital hardware design. This book addresses both theoretical and practical aspects in the design of a high-level synthesis system that transforms a behavioral level description of hardware to a synchronous logic-level implementation consisting of logic gates and registers.

The present level of maturity of synthesis techniques for digital circuits has been proven by the large number of designs achieved by means of computer-aided tools. Synthesis from behavioral descriptions can leverage the expressive power of high-level models and shorten the design time of a circuit. High-level optimization algorithms can be extremely powerful in making decisions about the macroscopic structure of a circuit, with beneficial effects on the overall performance, area and testability.

Synthesis techniques provide a competitive edge in the design of *Application Specific Integrated Circuits* (ASICs), which represent a growing sector of the semiconductor market. In particular, high-level synthesis can be extremely useful to ASIC design, because of the paramount importance of short time to market.

This book addresses specific issues in applying high-level synthesis techniques to the design of ASICs. This complements previous results achieved in synthesis of general-purpose and signal processors, where *data-path* design is of utmost importance. In contrast, ASIC designs are often characterized by complex *control* schemes, to support communication and synchronization with the environment. The combined design of efficient data-path and control-unit is the major contribution of this book.

Three requirements are important in modeling ASIC designs: *concurrency*, *external synchronization*, and *detailed timing constraints*. The objective of the research work presented here is to develop a hardware model incorporating these requirements as well as synthesis algorithms that operate on this hardware model.

The contributions of this book address both theory and implementation of algorithms for hardware synthesis. The following novel synthesis algorithms have been developed: *relative scheduling*, a scheduling formulation that supports external synchronizations and detailed timing constraints; *constrained conflict resolution*, a method to resolve resource conflicts under timing constraints; *relative control synthesis*, a control generation approach for relative scheduling; *relative control optimization*, a novel optimization strategy that minimizes the size of the control implementation under timing constraints.

In addition, we developed a hardware description language called *HardwareC* with particular attention to ASIC design specification issues. The system implementation of the algorithms is provided by programs *Hercules* and *Hebe*, which serve as the front-end to Stanford's Olympus Synthesis system. They have been applied to the synthesis of benchmark examples and to some chip designs at Stanford University.

D. C. Ku
G. De Micheli

Acknowledgements

The high-level synthesis research described in this book is part of an on-going synthesis research project at Stanford University called the *Olympus* synthesis system. We would like to acknowledge and thank the many people who have contributed to this research.

We would like to acknowledge the significant contributions of David Filo to the project. He co-developed the theory behind the control resynchronization optimization, described in Chapter 9 of this book. He implemented control optimization and register folding in program Hebe. Other supporting tools of the *Olympus* suite were developed by Frédéric Mailhot, who was responsible for the logic-level framework, which included technology mapping, logic simulation, and the SLIF interchange format and by Thomas Truong, who wrote a simulator for SIF and a graphic display package.

Rajesh Gupta spent a vast amount of time generating design and benchmark examples in HardwareC, working with Claudionor Coelho on the network co-processor and with Thomas Truong on the second version of the DAIO chip. Dave Kasle designed the MAMA decoder chip. Michiel Ligthart designed the first version of the DAIO chip. We are indebted to them for their tenacious work in using the prototype versions of the synthesis system. We also thank Polly Siegel, Jerry Yang and Maurizio Damiani for their suggestions and criticisms.

We would like to thank Daniel Weise and Teresa Meng for their comments and criticisms, and Joe Kostelec, Uzi Bar-Gadda and Rindert Schutten of Phillips Research Laboratory for their industrial perspectives. Most important of all, we would like to thank our families for their love, understanding, and encouragement. None of this would have been possible without their support.

This research was sponsored by a fellowship provided by Phillips/Signetics, by NSF/ARPA, under grant No. MIP 8719546, and by AT&T and DEC jointly with NSF, under a Presidential Young Investigator Award program.

Chapter 1

Introduction

The trend of *Very Large Scale Integration* (VLSI) circuit designs is towards greater density and complexity. A decade ago, the size of chip designs ranged from tens of thousands to hundreds of thousands of components. Today, they consist of millions of components. This increase in chip density means that complex designs with more functionality can now be implemented as a single chip. An effective way to deal with the increasing complexity of designs is to raise the level of abstraction at which circuits are designed. *High-level synthesis* refers to computer-aided design approaches starting from the algorithmic description level. The benefits of such a methodology include shortened design time to reduce design cost, ease of modification of the hardware specifications to enhance design reusability, and the ability to more effectively explore the different design tradeoffs between the area and performance of the resulting hardware.

Previous work in high-level synthesis addressed mainly general-purpose processor and signal processing designs [MPC90]. In these designs, the behavior usually consists of a set of computations that are performed within a certain amount of time. Synthesis of these designs can produce cost-effective implementations because the synthesis system can take advantage of domain-specific knowledge to optimize the underlying architecture. In contrast, *Application-Specific Integrated Circuit* (ASIC) designs perform computations that are *specific* to a particular application. An example is an Ethernet co-processor that coordinates the activities between a microprocessor and a Ethernet line. In this case the co-processor is constrained to both the microprocessor architecture and the Ethernet protocol. Complex ASIC designs require complicated handshaking protocols to interface between concurrently executing components and need to satisfy strict timing constraints on the handshaking. For these designs, synthesizing the control is as important as synthesizing the data-path, which has been traditionally the emphasis of high-level synthesis systems.

We believe ASIC designs to be particularly suited for high-level synthesis because the manual synthesis of these designs is tedious and error prone. The use of a high-level synthesis methodology significantly reduces the design time and cost, which is often as important as minimizing area or improving performance. However, existing synthesis approaches do not adequately address all the issues related to the synthesis of ASICs. For example, most systems cannot handle designs with detailed timing constraints; among the systems that do, even fewer can support complicated handshaking and interfacing requirements.

This book focuses on the issues that arise in constructing a high-level synthesis system for synchronous ASIC designs, and addresses both theory and implementation. The theory includes novel formulations and algorithms for hardware synthesis that uniformly incorporate *concurrency*, *synchronization*, and *detailed timing constraints* in the behavioral model. The algorithms have been implemented in a high-level synthesis system consisting of two programs called *Hercules* and *Hebe*. This system takes as input the *HardwareC* language, developed as a test-bed to experiment with language constructs for synthesis, and produces synchronous logic-level implementations of both the data-path and control. Hercules and Hebe have been applied successfully to the synthesis of benchmark circuits and chip designs at Stanford University.

1.1 Overview of High-level Synthesis

High-level synthesis takes a specification of the behavior of a digital system along with a set of constraints and goals on the resulting hardware to be satisfied, and finds a structure that realizes the given behavior while satisfying the given goals and constraints. The behavior is usually described as an *algorithm*, similar to programming language descriptions. The structure is a *register-transfer* level implementation that includes a *data-path* portion and a *control* portion. The data-path contains a network of functional units, registers, and their interconnection. The control activates components of the data-path to realize the required behavior. One of the objectives of synthesis is to find a structure that satisfies the constraints, such as requirements on area, latency or cycle-time, while minimizing other costs. For example, the goal might be to minimize the area while meeting timing requirements.

Due to its complexity, high-level synthesis is divided into a number of distinct yet inter-dependent tasks [MPC90]. To explain these tasks and their interrelationship, we describe a typical synthesis flow. First, a hardware description language specification is parsed into an internal representation that models both the control-flow and data-flow of the input behavior. The internal representation is optimized by compiler-like transformations such as dead-code elimination, common subexpression elimination and constant propagation. *Scheduling* and *binding* are then performed to map the behavior into structure. Scheduling assigns operations to control steps, where a control step is the fundamental unit of sequencing in synchronous systems and corresponds to a clock cycle. Binding assigns operations to specific allocated hardware resources. They are closely related and inter-dependent. For example, scheduling attempts to minimize the number of required control steps subject to the amount of available hardware which depends on the results of the binding. Likewise, binding exploit concurrency among operations to allow sharing of hardware resources, where the degree of concurrency is determined by scheduling.

For a given schedule and data-path, control logic must be synthesized to activate components in the data-path according to this schedule. There are many different ways of synthesizing the control, ranging from hardwired control, such as finite-state machines, to microcoded control, where each control step in the schedule corresponds to a microprogram instruction.

The major difficulty in high-level synthesis is the large number of design alternatives that must be examined in order to select the design that best meets the design goals and still satisfies the constraints. The computational complexity of even one of the above subtasks can be intractable in the presence of constraints. For example, the problem of scheduling under resource constraints is provably NP-complete [GJ79]. Compounding the problem is the tight inter-dependencies that exist between scheduling and binding that make the *design space* of possibilities multi-dimensional and irregular. For high-level synthesis to be practical, it is necessary to find an acceptable compromise between the degrees of freedom in exercising design choices and the complexity of the synthesis computation.

1.2 Motivation and Issues in High-level Synthesis

Logic synthesis techniques are well established and have been used for industrial ASIC chip designs [dG89]. In contrast, very few commercial designs have been synthesized using high-level synthesis techniques. This lack of acceptance is most likely due to a mismatch between the requirements of ASIC designs and the assumptions and capabilities of existing high-level synthesis systems. We

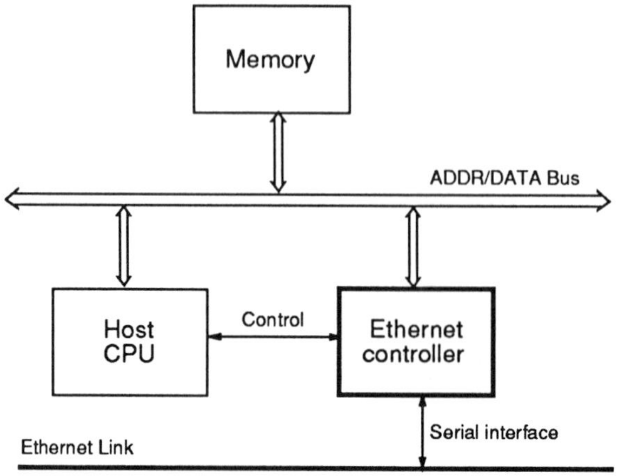

Figure 1.1: Block diagram of an Ethernet co-processor.

describe in Section 1.2.1 an Ethernet co-processor chip that highlights several important characteristics of ASIC designs. We then summarize these requirements in Section 1.2.2 as motivation for this research.

1.2.1 Example of an ASIC design

An example of a complex ASIC design is an *Ethernet co-processor*. The co-processor manages the processes of transmitting and receiving data frames over a network under CSMA/CD (Carrier Sense Multiple Access with Collision Detection) protocol [Tan81]. Its purpose is to off-load the host CPU from managing communication activities. The CPU, co-processor, and system memory all share an address/data bus where possession of the bus is ensured by handshaking signals between the modules. A block diagram of the co-processor is shown in Figure 1.1.

To receive data from the Ethernet line, the host CPU sets aside an adequate amount of buffer space and invokes the co-processor by writing to a memory mapped address. The co-processor responds by requesting bus control, and then reading the command instruction from memory. Upon enabling the co-processor in receive mode, the asynchronously arriving data frames are stored into a free memory area. Once an entire error-free frame is received, the co-processor fetches the adress of the next free receive buffer and interrupts the CPU. Transmission of data is performed in a similar manner.

Handshaking plays a crucial part in the specification and synthesis of this co-processor. For example, in writing the received data frames to memory, handshaking is required between the co-processor and host CPU to secure the bus access. Another example of handshaking can be seen in the implementation of the CSMA/CD protocol. Carrier sense means any station on the Ethernet wishing to transmit listens first to the line and defers its transmission until the channel is clear; collision detection refers to the collision detection protocol used to detect simultaneous transmission by two or more stations. Complicated handshaking is needed for error recovery and collision detection between the co-processor and the Ethernet line. In addition, the co-processor must satisfy timing constraints on the handshaking to ensure that real-time processing of the received data frames is possible. This general constraint can be translated into detailed timing requirements on individual co-processor operations. Timing constraints are also required by the request and acknowledge handshaking between the co-processor and host CPU for bus arbitration.

1.2.2 Requirements of ASIC synthesis

A major reason for the lack of acceptance of high-level synthesis is due to the difficulty of integrating a synthesized design with other components in the system. System integration constrains the *interface* of a synthesized design [Bor88b]. The interface takes into account environmental factors under which the chip will be operating. In particular, a synthesized design needs to communicate with other modules in the system using a given handshaking protocol and possibly under timing requirements. For example, the network co-processor design in the previous section needs to both communicate with the Ethernet line under the CSMA/CD protocol and handshake with the host CPU under a given bus arbitration protocol.

Existing synthesis approaches and algorithms are limited in their ability to synthesize designs with these complex interface and communication requirements, with a few exceptions [BK87, NT86, HPG88]. In particular, the input languages of most synthesis systems do not support synchronization of designs with input signals and events. Furthermore, the synthesis paradigm of most systems only considers the overall latency between the inputs and outputs of the synthesized design. Designs with complex handshaking protocols often require complex timing specifications for communication with the other modules in the system. Therefore, two important issues in the synthesis of ASIC designs are the support for *external synchronization* with input signals and events, and *detailed timing constraints* that specify bounds on the timing of specific input and output events.

In addition to handshaking and timing constraints, ASIC designs require the design system to support synthesis with partial structures and links to logic synthesis. Designers often have a specific structure or architecture in mind for a portion of their design. For example, the designer may explicitly specify that an operation be implemented by a particular functional unit in the final implementation, or that a group of operations be implemented by the same functional unit in order to manually control the degree of hardware sharing. Such *partial structure* information must be incorporated as part of the synthesis process [Kna89]. This information can be used to limit the number of different design alternatives during synthesis, to speed up the synthesis process.

Combinational and sequential logic optimization and synthesis techniques are becoming mature and have wide applications in both industry and research [dG89]. However, they have not been adequately integrated with high-level synthesis approaches. More specifically, most existing synthesis systems assume that the primitive building blocks of the synthesized hardware consist of predefined architectural templates. A 4-input multiplexer and an 8-bit adder are examples of these primitive components. For the synthesis of ASIC designs, this can result in inflexible design styles and an inability to take advantage of logic synthesis techniques because a large portion of an ASIC's functionality is described as combinational logic expressions. Therefore, logic synthesis is useful to both minimize the expressions and extract accurate delay and area estimates to guide high-level design decisions. Since the controller is a significant portion of most ASIC designs, FSM and logic synthesis techniques can improve the quality of the resulting control implementation.

1.3 Research Objectives

We identify the following three important requirements in modeling ASIC designs: concurrency, synchronization, and detailed timing constraints.

- *Concurrency.* Hardware systems are inherently parallel. This parallelism needs to be modeled so synthesis can explore tradeoffs between serial and parallel designs.

- *Synchronization constraints.* Concurrently executing modules need to synchronize with one another to coordinate their actions. Synchronization is especially important if the design is to communicate with its environment using a given handshaking protocol. Since when an external synchronization occurs is not known *a priori*, it can be represented in the hardware model as an operation with *data-dependent execution delay*. For example, detecting the rising edge of an input signal is a data-dependent delay operation because we do not know statically how long it will take before the signal is asserted.

- *Detailed timing constraints.* These constraints specify strict timing relationships, represented as minimum and maximum bounds on the activation of operations, that affect the externally visible behavior of a given module. For example, we may require the Ethernet co-processor to read the address/data bus at most 8 cycles after the rising edge of a request signal, and consecutive writes to a bus should take place at most once every 2 cycles.

We describe in this book a hardware model that incorporates these three requirements, and develop synthesis algorithms that operate on this new hardware model. The algorithms guarantee that a synthesized design satisfies its synchronization and timing requirements, or indicate when no such solution exists.

1.3.1 Organization of synthesis system

A hardware description language called *HardwareC* was developed to capture the three requirements listed above. HardwareC is the input language to the *Hercules* and *Hebe* high-level synthesis system that is the system implementation of algorithms developed in this research[1].

A block diagram of the synthesis system is shown in Figure 1.2. The input HardwareC description is parsed and optimized by Hercules using compiler optimizations to produce a sequencing graph abstraction of hardware behavior. Hebe maps the sequencing graph to a synchronous logic-level implementation consisting of logic gates and registers.

Traditionally, synthesis algorithms have focused on *resource-dominated* circuits, such as digital signal processing and processor-based designs. In such a design, the overall area and latency are primarily a function of the area and latency of its data-path components (e.g., functional units and registers). An appropriate synthesis strategy for this paradigm is to *schedule under resource*

---

[1] Hercules performs the dirty work such as parsing HardwareC and cleaning stables. His wife, Hebe, performs the core synthesis tasks such as scheduling and binding, and also serves wine to the gods on Mount Olympus.

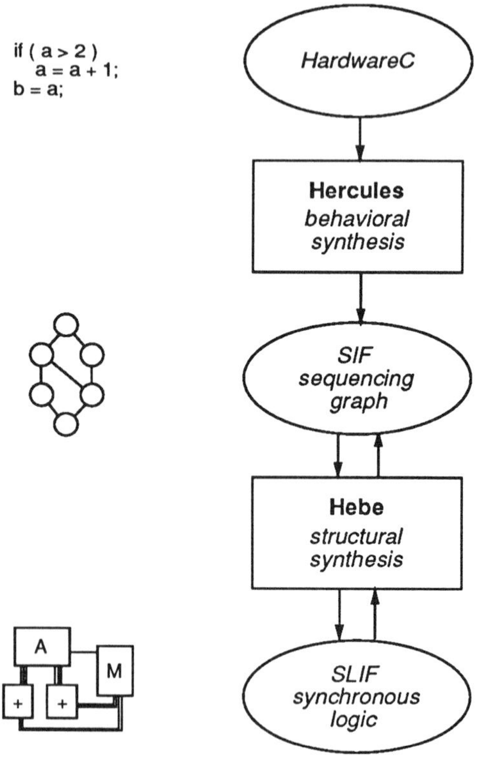

Figure 1.2: Hercules and Hebe high-level synthesis system.

constraints, since reducing the total number of data-path components most likely results in a corresponding decrease in the overall implementation cost. The control portion for these designs is often negligible.

In contrast, the control portion is a significant part of the overall hardware cost in *control-dominated* ASIC circuits. For these designs to operate correctly, they typically must satisfy strict synchronization and timing constraints on their interfaces to limit the range of possible I/O behaviors in the resulting implementation. Since the control and multiplexing costs are significant factors in the area and latency for these designs, applying the same synthesis strategy as for resource-dominated designs may yield inferior implementations because the assumptions underlying the optimization criteria may be incorrect. For example, the multiplexer delay into a heavily shared resource may be longer than the control cycle time, thereby invaliding the schedule.

Algorithms have been proposed to perform scheduling together with binding. However, their algorithmic complexity usually limits their applicability for non-trivial designs. To manage the complexity, our synthesis strategy performs resource binding *before* scheduling. More specifically,

1. *Perform resource binding under resource constraints.* We formulate the *design space* independently of scheduling, by defining the design space to be the different ways of binding abstract operations to allocated functional units. The designer is guided through the design space by both *exact* and *heuristic* search strategies.

2. *Perform scheduling under detailed timing and synchronization constraints.* The resulting schedule is used to synthesize the control unit.

This strategy gives the designer fine-grain control over the possible design choices. Furthermore, partial binding information can easily be incorporated to limit the size of the design space. This approach is especially appropriate for ASIC designs because the number of shareable functional units in these designs is usually limited. Therefore, for a particular binding, synthesis performs the difficult task of producing a control implementation that satisfies the synchronization and timing requirements.

Logic synthesis techniques are used to provide more accurate estimates on the delay and area of these modules to drive the scheduling phase. The use of logic synthesis is in contrast to many synthesis systems that use predefined microarchitectural components whose delay and area information is characterized and stored as attributes in a library. For ASIC designs, using predefined components may lead to inferior logic-level implementations.

1.3.2 New synthesis algorithms

Novel synthesis formulation and algorithms developed as part of this research include the following:

- *Relative scheduling:* a scheduling formulation that supports synchronizations and timing constraints,

- *Constrained conflict resolution:* a method to resolve resource conflicts under timing constraints,

- *Relative control synthesis:* a control generation approach that synthesizes control circuits from a relative schedule as an interconnection of interacting finite-state machines, and

- *Relative control optimization:* a technique, called *control resynchronization*, that reduces the control implementation costs while still satisfying the required timing and synchronization constraints.

Various aspects of this research have been reported in literature [MK88] [KM90b] [KM90a] [KM90c] [MKMT90] [KM91b] [KM92] [KFM91] [KM91a] [CE91]. The major theoretical contributions of this research are briefly summarized below.

Relative scheduling formulation. Traditionally, scheduling assigns operations to control steps. The scheduling problem can be formulated as an integer labeling problem that assigns to each operation an integer value representing its time offset from the beginning of the schedule. However, when external communication and synchronization are part of the behavioral model, *data-dependent delay operations* invalidate the traditional scheduling formulation because an absolute schedule satisfying timing constraints no longer exists.

To synthesize a design with complex handshaking requirements, traditional systems usually decompose the problem into smaller components, with each component containing only fixed delay operations which can be scheduled independently. The control to coordinate these components subject to the handshaking requirements is manually generated. The main limitation of this approach is the difficulty of specifying and enforcing timing constraints on the handshaking requirements because operations are now scattered into separate modules. Furthermore, an otherwise integrated design specification is now decomposed into many small pieces, which may be unnatural for the designer. Finally, since most systems synthesize a single hardware module at a time, this decomposition can limit the degree of hardware sharing that is possible.

Alternatively, this research incorporates external and internal[2] synchronizations into a novel formulation called *relative scheduling*. Relative scheduling supports the scheduling of both fixed and data-dependent delay operations under detailed timing constraints. It provides a theoretical framework for analyzing consistency of timing constraints in the presence of data-dependent delays. Polynomial-time algorithms have been developed for such consistency analysis, and they have been efficiently incorporated into a vertically integrated synthesis framework. Details of these algorithms and analysis of their properties are presented in Chapter 6.

[2]For example, data-dependent while loops can be represented as internal synchronization points.

The relative scheduling formulation provides a theoretical basis for analyzing *redundancy* in the synchronization of a given operation. Using synchronization redundancy can reduce the size of the corresponding control circuit, and algorithms are presented to remove all redundancies in a schedule.

Constrained conflict resolution. Hardware components can be shared to reduce design area. Synthesis algorithms must ensure that no two operations sharing the same hardware unit can execute simultaneously by *resolving* these resource conflicts. Furthermore, timing constraints must not be violated by any hardware sharing. In traditional approaches that consider only fixed delay operations, conflict resolution is combined with scheduling to ensure no hardware sharing between any two operations assigned to the same control step. This strategy must be extended to support the presence of data-dependent delay operations in our hardware model since static assignment of operations to control steps does not in general exist.

This research formulates the problem of conflict resolution as the task of *serializing* operations bound to the same hardware resource. A technique called *constrained conflict resolution* was developed to resolve resource conflicts such that the resulting design still satisfies the required timing constraints. This technique uses the constraint topology to reduce its computational complexity. Satisfiability of timing constraints is based on relative scheduling. Therefore, the approach is guaranteed to resolve resource conflicts such that timing constraints are satisfied, or indicate when no such solution exists.

Relative control synthesis. Relative scheduling complicates the task of control generation. When there are no data-dependent delay operations, the schedule consists of a single sequence of control steps that can be synthesized by traditional control strategies. For example, the schedule can be implemented as a microprogrammed controller or as a single finite-state machine. In the general case, however, these traditional control schemes are inadequate.

A problem arises because the hardware behavior consists of a number of concurrent execution sequences, each of which may depend on independent external synchronization conditions. Consider for example two execution sequences, where one sequence involves a set of operations that executes after detecting the rising edge of a signal and the other sequence depends on detecting the rising edge of another signal. Control for each execution sequence can be implemented as a finite-state machine. However, these individual finite-state machines need to be coordinated to ensure that the overall control functions correctly. For designs with complicated handshaking and synchronization requirements, this integration is nontrivial.

An approach called *relative control synthesis* was developed to address these issues. In relative control, control logic is synthesized by directly mapping the sequencing graph model to an interconnection of finite-state machines (FSMs). For each external synchronization in the behavior, a modular FSM is generated to activate operations that depend on this synchronization. The FSMs interact with one another through a set of handshaking signals. The approach supports external synchronization, multiple threads of concurrent execution flow, and control generation from relative scheduling. In addition, the resulting control unit is guaranteed to yield the *minimum* number of cycles in executing the hardware behavior for all input data sequences. This feature is important to ensure that the control unit incurs no performance penalties for the arbitrary nesting of model calls, conditionals, and loops in the behavioral specification.

Relative control optimization by resynchronization. The control circuit resulting from control synthesis can be optimized further to reduce its size, with the requirement that the optimized hardware is still a valid implementation of its behavioral model. Most approaches to control optimization use a finite state machine model, where operations are bound to control states. However, when synthesizing circuits from a higher, more abstract level of hardware specification that supports concurrency, synchronization, and timing constraints, these approaches may be overly restrictive.

We present in this book a control optimization strategy based on *resynchronization of operations*. Using the relative control synthesis strategy, we show first how the graph can be mapped to a control architecture consisting of synchronization and offset control components. In contrast to previous work that focused on reducing the latency of the implementation, we assert that the minimal latency solution is not always needed. Instead, we focus on reducing the control area while simultaneously satisfying the timing constraints that can include upper bounds on latency. The result of this approach is a control implementation with potentially smaller area while still "fast enough" to meet the specification. We reduce the total control cost by introducing *synchronization redundancy* in the graph, where any modification to the graph is considered to be acceptable provided the original timing constraints are not violated.

1.4 Summary of Related Work

This section briefly contrasts our approach with the related research in high-level synthesis, broadly described by the tutorial article by McFarland, et. al. [MPC90], the books edited by Camposano and Wolf [CE91] and by Gajski [bDG88], as well as the survey book by Walker et. al. [WC91]. For in-depth description of the algorithms used by other systems, we refer the interested reader to the books by Gajski et. al. [GDWL92] and Gebotys et. al. [GE92].

1.4.1 Overview of existing approaches

The focus of most high-level synthesis efforts todate has been on synthesizing and optimizing the data-path. Three important tasks in data-path synthesis are *hardware allocation*, *module binding* and *scheduling* [MPC90]. Hardware allocation determines the number of functional units, registers, and multiplexers that are needed to synthesize a given behavior. It is strongly tied to module binding which assigns operations to specific hardware modules. Scheduling assigns operations to control steps. These tasks are closely related; the results of one directly affect the others. For example, whether two operations can be scheduled in the same control step depends upon whether they share common hardware resources, which in turn depends on hardware allocation and module binding. Computationally, finding the best hardware configuration under hardware and timing constraints is extremely difficult. Even the subproblems of scheduling under resource constraints and scheduling under precedence constraints with release times and deadlines are NP-complete [GJ79]. To manage the complexity, heuristic approaches to these tasks have been proposed. Existing synthesis systems can be categorized according to their interplay among the allocation, binding, and scheduling phases.

One common synthesis approach is to set a limit on the amount of available hardware and schedule subject to this limit. In this formulation, operations are first statically assigned to specific control steps using heuristics to minimize the critical path delay. Resources are shared as much as possible subject to the given schedule to reduce the area cost of the design. Examples of systems that adopt this methodology include the CMU-DA System at Carnegie-Mellon University [TIK+83], the Design Automation Assistant [Kow85], the Flamel system at Stanford [Tri87], and Facet [TS86]. Note that since the data-path has not yet been constructed, accurate delay estimates for the interconnect are not known during scheduling. This can lead to erroneous scheduling due to incorrect assumptions.

A more flexible synthesis approach is to *iterate* over these tasks, first allocating and binding under partial schedule, then scheduling under partial allocation until a satisfactory solution is obtained. This is done under user control in MIMOLA [Zim79] and under an expert system paradigm in Chippe [BG87]. The RLEXT tool [Kna89] allows a user to freely modify a data-path design. The modified data-path is automatically repaired to adhere to a given behavior. This ability supports tight interaction between a partial structure and its corresponding schedule.

To obtain accurate interconnection delays to guide scheduling, techniques such as bottom-up design synthesis in BUD [McF86] or inter-dependent scheduling and binding using priority list scheduling variants have been used [PPM86, PG87, PKG86]. Some approaches even model scheduling and allocation as a combinatorial optimization problem. They use techniques such as linear programming [BM89] and simulated annealing [DN89] to solve these problems. Although these approaches are effective in closing the gap between top-level design methodologies and bottom-level design information, existing implementations are limited due to the complexity of the combined problem.

An alternate approach performs hardware allocation and binding before scheduling [CR89, McF86]. In this formulation, operations are first bound to specific resource components. Scheduling is then applied to ensure that all operations bound to the same resource component are disjoint in time to avoid resource conflicts. In the Caddy synthesis system [CR89], for example, the designer first specifies the maximum resource utilization per control step prior to scheduling, then schedules using analysis tables to minimize the delay. Manual intervention is required to make tradeoffs between resource utilization alternatives. Timing constraints on sequences of operations are supported. Analysis of timing constraint consistency by successive graph reduction is possible because of the series-parallel nature of the control flow. However, distribution of a constraint over a sequence of operations is based on a heuristic scheme, which makes it difficult to rigorously analyze the consistency of timing constraints. This research extends this synthesis approach to address the issues of timing constraint consistency and design space exploration.

Other approaches model hardware systems by net-based representations, such as the Petri-net model and its derivatives [Pet83], and event based models [Bor88a]. Synthesis can be seen as a series of transformations on the net model so that the resulting net model satisfies certain properties necessary for hardware synthesis. Restrictions of Petri-nets have already been widely used for asynchronous circuit modeling and synthesis [Chu87, MMB89]. There is active research in using net-based models as the underlying model for analysis and synthesis [Rok90].

1.4.2 Critical review and research contributions

While the systems described in the previous section have been effective in synthesizing certain types of designs and efficient algorithms have been developed to address many difficult synthesis problems, existing approaches do not address the synthesis of ASIC designs with complex handshaking protocols and strict timing requirements.

Most approaches assume that the execution delay of operations is bounded, which stems from the use of pre-designed micro-architectural library modules as primitive hardware elements for the data-path. This implies that hardware interfacing and synchronization, modeled as operations with data-dependent execution delay, are not supported. In contrast, this research incorporates external interfacing and handshaking requirements as an integral part of hardware model and performs synthesis based on this hardware model. The support for handshaking is important in ASIC designs to coordinate the actions between a set of concurrently executing modules.

In contrast to micro-architectural synthesis approaches where the final implementation is an interconnection of primitive functional blocks, this research uses logic synthesis as the underlying synthesis base. The characterization of resources to evaluate hardware sharing feasibility is carried out using logic synthesis techniques to provide estimates on timing and area. This methodology is particularly suited for ASIC designs that tend to rely on application-specific logic functions. The use of logic synthesis for estimates improves the quality of the synthesized designs and avoids erroneous high-level decisions due to insufficient data or inappropriate assumptions.

The task of scheduling is central to a high-level synthesis strategy. With the exception of CALLAS, SAW, and SALSA [NK90], most synthesis approaches do not support detailed timing constraints. That is, they support either no timing constraints at all or they support at most constraints on the overall latency. This may be inadequate to describe complicated requirements on the timing of operations. SAW, because of the heuristic nature of its scheduling step, cannot guarantee that if the algorithm fails to find a solution that satisfies the timing constraints then no solution is possible. Rigorous analysis of the consistency of detailed timing constraints is either limited or lacking. Therefore, only a handful of systems perform synthesis under timing constraints, and among them, even fewer support analysis of the consistency of timing constraints. In contrast, this research considers synthesis under detailed timing constraints in both the synthesis formulation and algorithms.

In summary, this research differs from other synthesis approaches by uniformly incorporating concurrency, external synchronization, and detailed timing constraints into the hardware model. In particular, the relative scheduling formulation serves as the theoretical basis for analyzing the consistency of detailed timing constraints in the presence of data-dependent delay operations. The proposed synthesis algorithms presented in the sequel guarantee that these synchronization and timing requirements are satisfied by the resulting synthesized hardware.

1.5 Assumptions and Limitations

In this book, we assume a *synchronous* and *non-pipelined* hardware implementation of the synthesized design. The clocking methodology is synchronous *single-phase* where delay elements in the resulting hardware are implemented as edge-triggered flip-flops. Synthesis of multi-phase synchronous designs is not considered in this research. In addition, both the control and data-path portions of the implementation operate off a single phase.

The second assumption of non-pipelined implementation implies that given an input behavior consisting of an algorithmic flow of computation, the last operation must complete before reactivating the first operation in the algorithm flow. In other words, an operation in the behavior is executed at most once for each execution of the behavior. Pipelined designs can be supported explicitly by describing each pipe stage as an independent hardware process, synthesizing each individual process independently, then connecting the stages together to form the pipeline.

Based on the above assumptions, the term "high-level synthesis" in the remainder of this book refers to the high-level synthesis of synchronous, non-pipelined digital circuits.

1.6 Outline of the Book

After presenting the objectives of this research and its relationship with existing synthesis approaches in the introductory Chapter 1, Chapters 2 through 5 describe the hardware model and synthesis framework for *Hercules* and *Hebe*. They set the context in which the novel synthesis algorithms presented in subsequent chapters are applied. More specifically, Chapter 2 presents an overview to the overall synthesis framework and describes the modeling of hardware using the HardwareC language. Chapter 3 describes the behavioral transformations performed by Hercules to identify parallelism in the input description that include parsing and compiler optimizations. Chapter 4 describes the *sequencing graph* model, which is the underlying representation for all synthesis formulation and algorithms. The treatment of hardware resources is also described in that chapter. Chapter 5 describes the formulation, construction, and exploration of the design space in Hebe.

The main algorithmic contributions of this research are described in the next four chapters. Chapter 6 presents the relative scheduling formulation that includes description of the algorithms and analysis of their properties. Chapter 7 describes conflict resolution under timing constraints. Chapter 8 describes the generation of the control circuit from a relative schedule. Chapter 9 describes the control resynchronization optimization that reduces the area of the control implementation under timing and synchronization constraints.

Finally, the system implementation of Hercules and Hebe is described in Chapter 10. Chapter 11 describes several design scenarios and presents the experimental results of applying the synthesis system to the design of benchmark circuits and chip designs. Chapter 12 concludes this research and proposes avenues of further research.

Chapter 2

System Overview

We overview in this chapter the hardware modeling language and synthesis flow in Hercules and Hebe. Section 2.1 describes the modeling of hardware behavior using *HardwareC* [KM90a]. HardwareC supports constraint specification and external synchronizations. Section 2.2 presents a brief overview of the overall synthesis flow in Hercules and Hebe.

2.1 Modeling Hardware Behavior

The input to the synthesis system is a description of hardware behavior written in a high-level hardware description language (HDL) called *HardwareC* [KM90a]. The motivation for choosing HardwareC over other hardware description languages is that it allowed experimentation with different language constructs for synthesis. The interaction between specification and synthesis provides an effective framework for testing new synthesis approaches and algorithms.

Many HDLs have been proposed and used in both academia and in industry. Most of these languages are oriented towards hardware simulation and documentation. Notable examples include VHDL [LSU89] and Verilog [TM91]. Conversely, HardwareC targets the description of synchronous, synthesizable digital hardware. HardwareC addresses the following issues that are relevant for ASIC designs:

- *Concurrency and synchronization:* HardwareC describes complex circuits as a collection of concurrently executing processes. Communication and synchronization among these processes are supported by parameter passing and message passing mechanisms. Within a process, operations can execute either serially, in parallel subject to data dependencies, or in parallel unconditionally.

- *Timing and resource constraints:* Minimum and maximum timing constraints between the activation of operations can be explicitly specified in HardwareC. The language also supports resource constraints in terms of partial binding information and resource usage bounds.

- *Flexibility in design styles:* Hardware description languages capture an abstract design from the designer and should therefore support a variety of design styles, ranging from a pure behavioral description, to a mixture of behavior and structure, to a pure structural description of the design, including specific instantiations of hardware modules and their interconnection. HardwareC supports procedural semantics to describe algorithms and declarative semantics to describe structural interconnection.

As its name suggests, HardwareC has a C-like syntax. However, the language has its own hardware semantics and differs from the C programming language in many respects. The language allows the designer to specify *constraints* on the timing and resource requirements of a given design. Constraints are important for synthesis-oriented languages because they can guide the synthesis system in exploring implementation tradeoffs. We describe the major characteristics of HardwareC in this section. Details of the language are given in [KM90a].

Assumptions. HardwareC supports specific constructs for the design of *synchronous* digital circuits. In particular, HardwareC supports I/O, message passing, and register loading operations which are synchronized to a clock cycle. Similarly, data-dependent loops and process constructs have an underlying synchronous model, i.e. they require an integral number of clock cycles (possibly zero) to execute.

HardwareC assumes that the design is *non-pipelined.* In executing a process the control flow propagates from the first statement in the process to the last statement. Non-pipelined implementation means that the first operation in the process can be re-executed only once the last operation in the process has completed in the previous execution. Each time process is restarted, a statement in the process is executed at most once. Although the language does not directly support pipelined descriptions, each pipe-stage can be described as a separate process, where the interconnection of these pipe-stage processes forms the pipeline.

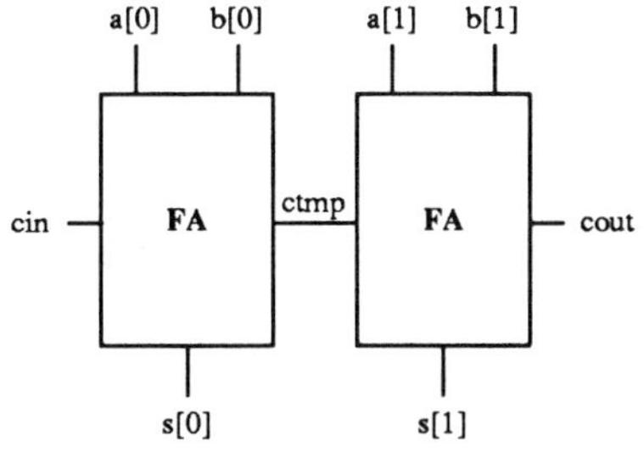

Structural interconnection
(a)

```
block ripple(a,b,s,cin,cout)
  in port a[2],b[2],cin;
  out port s[2],cout;
<
  boolean ctmp;
  FA(a[0],b[0],s[0],cin,ctmp);
  FA(a[1],b[1],s[1],ctmp,cout);
>
```

Block model description
(b)

Figure 2.1: Example of using block models to describe a ripple chain of adders in HardwareC.

2.1.1 Models

In HardwareC, hardware is modeled as a collection of concurrent modules. Each module represents functionality that can be described either as a structural interconnection of components (i.e. *declarative* semantics) or as a set of operations sequenced in time that performs a particular algorithm (i.e. *imperative* or procedural semantics).

There are four fundamental design abstractions in HardwareC: *block*, *process*, *procedure*, and *function models*. At the top level, a design is described as a block. A block describes the structural relationship and physical connectivity between different components of a design. It has a *declarative* semantic and describes an interconnection of logic and instances of other blocks and processes. For example, a block model that describes a ripple chain of adders is shown in Figure 2.1.

In contrast, process, procedure, and function models have an imperative semantic. An imperative semantic model describes an algorithmic flow of operations sequenced in time. The algorithm consists of operations such as logic expressions and assignments to shared variables, and control-flow constructs such as sequencing, branching, and iteration. Although process, procedure, and function models all describe an encapsulation of operations in the form of an algorithm, a *process* model in HardwareC differs from the others in that it repeatedly executes the algorithm, i.e. a process automatically restarts execution

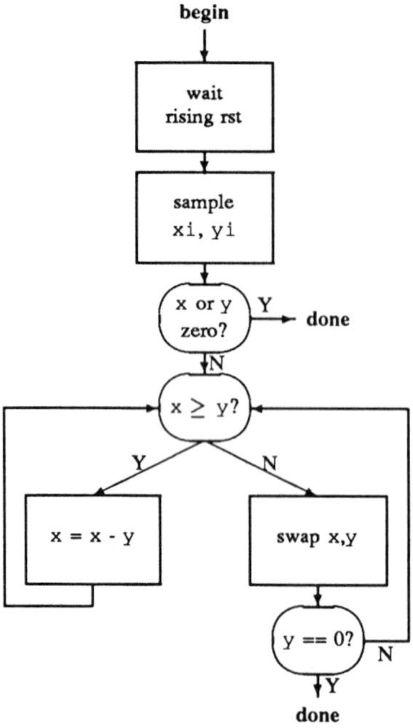

Flow chart of algorithm
(a)

```
process gcd(xi,yi,rst,result)
  in port  xi[8], yi[8];
  in port  rst;
  out port  result[8];
[
  static x[8] = 0, y[8] = 0;
  /* set output to 0 during algorithm */
  write result = 0;
  /* wait until rising edge of rst */
  while ( ! rst )
    ;
  /* sample input */
  {
    x = read(xi);
    y = read(yi);
  }
  if ((x != 0) & (y != 0)) {
    /* using Euclid's gcd algorithm */
    repeat {
      while (x >= y)
        x = x - y;
      /* x should be less than y now */
      /* so exchange x and y */
      < x = y; y = x; >
    } until ( y == 0 );
  }
  write result = x;
]
```

HardwareC process model
(b)

Figure 2.2: Example of a HardwareC process that repeatedly samples the inputs on the rising edge of `rst`, then finds the greatest common divisor of two input values using Euclid's algorithm.

upon the completion of its last operation. In contrast to a process, computation described by a *procedure* or *function* model executes when the model is *called.* An example of a process that finds the greatest common divisor of two numbers is shown in Figure 2.2. HardwareC supports only non-recursive procedures and functions.

Modeling hardware behavior as a collection of concurrent and interacting processes is natural for hardware description since hardware modules continuously operate on a time varying set of inputs. Therefore, *blocks* describe the structural relationships among the processes, which in turn describe algorithms containing a hierarchy of procedures and functions. Processes allow *coarse-grain* parallelism to be specified at the functional level.

Interprocess Communication The synchronization and data transfer between concurrently executing processes take place on *ports* that interconnect the processes. A process places the result of its computation on a given port, which can then be used as the input to another process by connecting the two ports together. In addition to ports, synchronization between processes can take place through *channels* on which send/receive message passing is performed. The two mechanisms for interprocess communication, *port passing* and *message passing*, are illustrated in Figure 2.3.

Port passing assumes the existence of a shared medium, such as wires or memory, that interconnects the hardware modules implementing the models. The protocol governing correct handshaking between the modules is provided by the designer and is described as an integral part of the high-level description. The ports are further categorized into "global" and "local" ports.

- *Global port* – Any access to a global port is *immediately* reflected to the other connected models. Global ports serve as linkages among block and process models. For procedures and functions, they allow direct access to the external signals, regardless of the nesting depth of the calling hierarchy, i.e. the value of a global port may change during the execution of a given model.

- *Local port* – In contrast, a local port transfers data across the calling hierarchy only. They serve as values that are passed in and out of procedure and function calls, and are valid only at the call boundary.

Message passing via explicit blocking *send* and *receive* operations can be used for both synchronization and data-transfer. Information transfer takes place

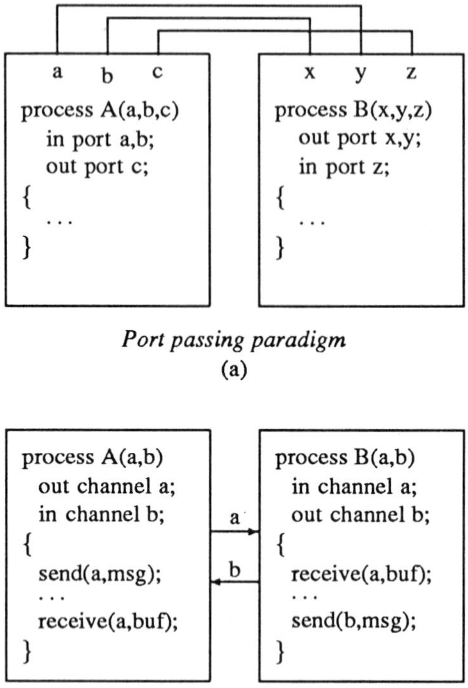

Port passing paradigm
(a)

Message passing paradigm
(b)

Figure 2.3: Port passing versus message passing mechanisms in HardwareC.

on *channels* that interconnect the models. The corresponding hardware for communication, as well as its protocol, are automatically synthesized by the synthesis tools under the assumption that the processes are synchronized to the same clock.

Block and process models can only have global ports and channels. For procedure and function models, both local and global ports are allowed in addition to channels.

2.1.2 Constants and variables

There are two major classes of data entities in the language – *constants* and *variables*. Constants can be decimal, binary (prefixed by and "0b"), and hexadecimal (prefixed by "0x"). Negative constants are represented using two's

Type	*Initial value*	*Implementation*
int	0	not synthesized (resolved at compile time)
boolean	0	wire or register
static	Previous call	register

Table 2.1: Variable types in HardwareC.

complement convention, with MSB on the left and LSB on the right, e.g., 0x1f is equal to 31, and 0b011 is equal to 3.

A variable is used to access the results of computation in a given description. There are three major variable types in the language: **int**, **boolean**, and **static**. An **int** variable can assume any integer value. They are mainly used in fixed-iteration loops as indices, and in accessing elements of Boolean vectors. Integer variables are *never* synthesized into hardware because they are always resolved at compile time. They are provided mainly for the designer's convenience. To emphasize the point of not being mapped to hardware, integer variables are often referred to as meta-variables.

A **boolean** variable represents one or more signals. Each bit of a Boolean variable corresponds to a signal that can have value 0 or 1. The total number of bits is the *size* of the variable. If the size is one, then it is a *scalar*; otherwise, it is a *vector*. An integer value is represented bit-wise using 2's complement convention; therefore, the acceptable values for a vector of size n range from $-2^{n/2} \ldots (2^{n/2} - 1)$. The indices of a vector start from 0 to $n - 1$, where index 0 represents the least significant bit (LSB), and index $n - 1$ represents the most significant bit (MSB).

A **boolean** variable is initialized to zero. Furthermore, the value assumed by the variable defined in a given model will not be retained the next time the model is invoked (for process models, the value is not retained when the process restarts). Depending on the decisions of the synthesis system, a **boolean** variable may be synthesized either as a wire or as a register in the final implementation.

Static variables are similar to **boolean** variables, with the semantic difference that their values are retained across procedural invocations. Since **static** variables have state information, they are always implemented as registers in the resulting hardware. Static variables may be optionally initialized to a pre-specified value upon system reset. The registers implementing static variables can be explicitly loaded by using the **load** assignment statement. For example,

Control-flow construct	*Syntax*
If-then-else	**if** (*cond*) *stmt1* **else** *stmt2*

Switch-case	**switch** (*cond*) {
	case *c1*: *stmt1* **break;**
	...
	default: *stmtk* **break;**
	}

For loop	**for** *i* = *c1* **to** *c2* **step** *c3* **do** *stmt*
While-do loop	**while** (*cond*) *stmt*
Repeat-until loop	**repeat** *stmt* **until** (*cond*)
Do-while loop	**do** *stmt* **while** (*cond*)

Table 2.2: Control-flow constructs in HardwareC.

```
/* loads register for status with 0xff */
load status = 0xff;
```

updates the register implementing `status` with the value `0xff`. Table 2.1 summarizes the variable types.

2.1.3 Language features

HardwareC supports a single-in, single-out control flow. This implies that no *gotos*, *breaks* from loops, and *returns* from procedures (other than at the very end) are allowed. It supports the usual iterative and branching control-flow constructs. Iteration can be specified either as fixed-bound loops (**for** loop) or data-dependent loops (**while** or **repeat-until** loops), depending on whether the loop exit condition is known at compile time.

Data-dependent loops can be used to detect signal transitions, e.g. the while statement in the *gcd* example of Figure 2.2 detects the rising edge of the `rst` signal:

```
/* wait until rising edge of rst (rst = 1) */
while ( !rst )
    ;
```

Table 2.2 summarizes the control flow constructs in HardwareC.

```
{
    x = y;
    y = x;
}
```

incorrect
(a)

```
{
    boolean temp;
    temp = x;
    x = y;
    y = temp;
}
```

using data-parallel
(b)

```
<
    x = y;
    y = x;
>
```

using parallel
(c)

Figure 2.4: Illustrating the differences between compound statements in HardwareC. (a) incorrect swap routine, (b) swap using data-parallel compound statement, and (c) swap using parallel compound statement.

We summarizes in the following sections major language features in HardwareC, with an emphasis on describing constructs that are useful for the specification of ASIC designs.

Varying degrees of parallelism

For imperative semantic models, HardwareC can describe varying degrees of parallelism in a given design through the use of *sequential*, *data-parallel*, and *parallel* compound statements, delimited by [ ], { }, and < >, respectively. In the first case, all operations in the compound statement are executed sequentially. In the second case, all operations are executed in parallel, unless data dependency requires serialization. In the last case, all operations unconditionally execute in parallel. Sequential grouping of operations is useful to describe explicit sequencing dependencies between operations, e.g. to detect the rising edge of a signal before reading from a given port.

Differences between these three groupings are illustrated by the following example. Consider the swapping of two Boolean variables x and y using the code segments of Figures 2.4(a), (b), and (c). The code segment in Figure 2.4(a) will *not* perform the swap because the first assignment transfers the value of y to x, so by the time the second assignment is executed, x already assumes the value of y and hence the second assignment is equivalent to assigning y onto itself. The reason lies in the data-dependency that exists between the two

```
template function ripple_adder(op1,op2,cin)
    with (size) return boolean[size]
  in boolean   op1[size], op2[size], cin;
{
  int   i;
  boolean   carry, result[size];

  /* ripple carry computation */
  carry = cin;
  for i = 0 to size-1 do
  {
    result[i] = op1[i] xor op2[i] xor carry;
    carry = op1[i] & op2[i] | carry & ( op1[i] | op2[i] );
  }
  return_value = result;
}
```

Figure 2.5: Example of a template description of a ripple carry adder function in HardwareC.

assignments that forces them to be executed in series. This error is corrected by introducing an intermediate variable `temp` to temporarily hold the swapped value in Figure 2.4(b). Finally, the parallel grouping in Figure 2.4(c) guarantees that the two assignments will be carried out in parallel and the right-hand sides of the assignments refer to values of variables *before* entering the compound statement. Therefore, the assignment `x = y` is semantically equivalent to `x` = *previous value of* `y`; similarly, the assignment `y = x` is equivalent to `y` = *previous value of* `x`.

Template models

Templates are models that take one or more integer arguments. They support *polymorphism* in the language by modeling several behaviors with a single description. For example, a single template can describe a family of adders of different sizes. Templates are therefore very useful in describing libraries of hardware operators at a high level. Templates are similar to generic packages in ADA or generic classes in several object oriented languages.

Given a particular mapping of integer values to integer parameters of a template model, a corresponding *instance* of the model is obtained. An analogy can be drawn between templates and module generation, i.e. a template is a form of high-level module generation. Consider for example the description of a template for the ripple-carry adder function in Figure 2.5. The template model `ripple_adder` takes an integer parameter `size` representing the operand size.

Explicit instantiation of model calls

A procedure or function model may be *called* by another model. The call indicates a request to execute the functionality defined by the called model, which is implemented by a particular *instance* of the hardware block corresponding to the invoked model. Therefore, from the standpoint of synthesis, a hardware *resource* is a model that can be shared in the hardware implementation. Resource sharing in high-level synthesis can be constrained by explicitly *binding* a call to a specific instance of the implementation.

There are two types of model calls: *hardware-unbound* and *hardware-bound* calls. A hardware-unbound call is a call where the particular *instance* of the called model used to implement the call is not specified. The synthesis system has the freedom to implement multiple hardware-unbound calls with one or more hardware resources. In some cases, however, the designer may wish to invoke a specific instance of the called model in the final implementation. This can be achieved by explicitly declaring an **instance** of the called model, similar to variable declarations. This hardware-bound call then invokes the declared *instance* of the model by binding the call in the resulting hardware.

There are several important advantages to explicitly instantiating models. First, the designer can access not only behavior through model calls, but also *internal state information* as well. This is analogous to the capabilities of abstract data type languages in software. Second, because a model instantiation is similar to instantiating a hardware module, *resource sharing* can be specified at the language level. Finally, the designer can *completely* specify the behavior that is intended without relying on hidden assumptions.

To illustrate the use of hardware-bound calls and explicit instantiation, consider the code segment of Figure 2.6(a) which increments or decrements the value `a` according to the value of `upload`. The code segment contains three hardware-unbound calls to the 5-bit `ripple_adder` template described in Figure 2.5. Initially, `a` is assigned to the sum of two variables `x` and `y`. In the most simplistic case, the choice of sharing the `ripple_adder` resource is decided by the synthesis system. However, suppose the designer wants to use only *a single* adder resource in implementing the description. To do so, the designer can explicitly declare an instance of the adder template, call it `INST`, and make

```
/* initially load */
a = ripple_adder(x,y,0) with (5);

/* loop and count */
while ( !reset ) {
   if ( upload )
      a = ripple_adder(a,1,0) with (5);
   else
      a = ripple_adder(a,-1,0) with (5);
}
```

Hardware-unbound model calls
(a)

```
/* declare instance */
instance ripple_adder with (5)  INST;

/* initially load */
a = INST(x,y,0);

/* loop and count */
while ( !reset ) {
   if ( upload )
      a = INST(a,1,0);
   else
      a = INST(a,-1,0);
}
```

Hardware-bound model calls
(b)

Figure 2.6: Example of (a) hardware-unbound model calls and (b) hardware-bound model calls in HardwareC.

all calls refer to that instance. In the case of Figure 2.6(b), the designer has achieved resource sharing at the functional level. Note that the designer can declare any number of instances and their bindings to calls. For example, the designer can declare two instances of `ripple_adder`, bind the first call to one instance and bind the remaining calls to the second instance.

Timing and resource constraints

HardwareC supports two categories of design constraints: *timing* and *resource* constraints. Timing constraints define upper and lower bounds on the time separation between operations. Resource constraints specify the number of resource components available, and partially bind calls to specific instances in the hardware implementation. The designer can also imbed arbitrary constraints in the description. This capability allows the designer to convey information that may be used by later synthesis steps.

Timing constraints can describe both *minimum* and *maximum* time separation between any two statements. Any statement in the language can be *tagged* by prefixing the statement with the tag name, followed by a colon (:) as delimiter. Timing constraints are specified with respect to the tags. Tags can be scalars or vectors and must be explicitly declared. Timing constraints are important in specifying timing relationships for a given protocol. An illustration of the use

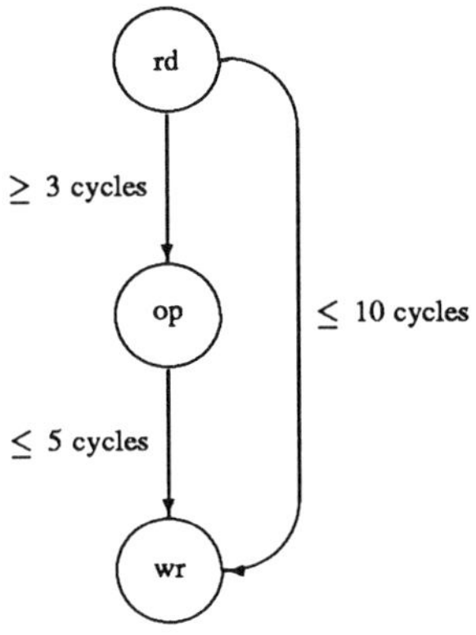

Constraint relationships
(a)

```
{
    tag  rd, wr, op;

    /* perform tasks */
    rd: data = read(input_port);
    op: result = some_function(data);
    wr: write output_port = result;

    /* specify timing constraints */
    constraint mintime from rd
          to op = 3 cycles;
    constraint maxtime from op
          to wr = 5 cycles;
    constraint maxtime from rd
          to wr = 10 cycles;
}
```

HardwareC code segment
(b)

Figure 2.7: Example of timing constraint specification in HardwareC.

of timing constraints is shown in Figure 2.7.

Resource constraints limit the number of instances that can be allocated for a specified type of resource. This upper bound constraint is described as attributes in the language. Partial binding of operations to resources using explicit model instantiation is another form of resource constraint.

2.1.4 Timing semantics

We describe now the timing semantics of the HardwareC language. Under the synchronous hardware paradigm, timing semantics define the number of cycles that is required to execute a particular statement in the language. Before we present the details of the timing semantics, we describe first the relationship between *values* and *assignments* for a given variable.

Assignments and values

For most register-transfer level hardware description languages used for synthesis, an assignment to a variable across basic block boundaries typically corresponds to loading the register implementing the variable with the assigned value. For example, an ISPS segment of an assignment to variable a is given below.

```
BEGIN
  a = 1 NEXT
  IF ( cond )
    a = 0 NEXT
  outport = a
END
```

This implies that a is loaded with the value 1 in the first control state; in the second control state, if `cond` is true, then a is loaded with the value 0.

This approach has the advantage of knowing exactly the execution delay for an assignment – one control state – and hence one can derive the execution delay for an entire model in a straightforward way. However, this advantage also leads to one of its principle drawbacks: the latency of a given hardware model is linked to its description style. For example, to evaluate a large expression, one style is to express it as a huge, monolithic expression; alternatively, one can express the expression as a set of partial computations. However, each partial computation will add an extra control state to the overall execution delay.

In contrast, HardwareC decouples the semantic of variable assignments from the loading of registers. When a program references a particular variable, it may reference different *values*, depending on whether the variable has been modified between the references. A variable is modified if it is the target of an assignment, or if it is used as the output parameter in a procedure call. Consider the following HardwareC segment in which a static variable v is assigned five times, where `expri` represents any expression yielding a value.

```
{
    v = expr1;              /* 1st assignment */
    v = expr2;              /* 2nd assignment */
    v = expr3;              /* 3rd assignment */
    write oport = v;
    v = expr4;              /* 4th assignment */
    write oport = v;
    v = expr5;              /* 5th assignment */
    write oport = v;
}
```

By replacing each of the three references to v with the corresponding last assigned value, we can resolve and remove the assignments to v. The resolved program segment is given below.

```
{
    write oport = expr3;
    write oport = expr4;
    write oport = expr5;
}
```

Since an assignment is always resolved and removed, it is not implemented by loading registers and hence does not require a clock cycle to execute. It is still possible, however, to explicitly load the implementing register by using the **load** construct for all the assignments[1], i.e., by prepending the keyword **load** before all assignments, in this case, all the assignments will require a control state to execute.

We explain the timing semantic of assignments in the following two steps:

1. *Resolution of all non-explicit load assignments.* First, all non-explicit load assignments are resolved and removed from the description by propagating the assigned value to the future references, as illustrated in the previous example.

2. *Introduction of selectively explicit load assignments.* For straight-line code segments, propagating the assigned values and removing the assignments is enough to ensure that a variable retains its value until its next assignment. However, in the presence of data-dependent loops, it is necessary to make explicit the loading of values into registers before a loop is entered or restarted. We concentrate on data-dependent loops only because fixed-iteration loops are always unrolled in our paradigm.

Let v be a static variable and reg_v be its implementing register. We identify three possible scenarios in which the *value* of a variable needs to be explicitly loaded into its implementing register. Even though the load statement may be omitted in a HardwareC description in the following cases, it is implied and therefore the corresponding statement will take one control state to execute.

- *Before entering into a data-dependent loop.* If v is modified or referenced in the body of a loop, then any value it assumes before entering the loop must be loaded into reg_v, so that the first reference to v in the body yields the correct initial value. Consider the following example.

[1] We consider only **static** variables because (1) **int** variables are always resolved at compile-time, and (2) **boolean** variables are either converted to **static** variables during behavioral optimizations, or they satisfy the single assignment requirement and are always implemented as wires.

```
v = 0;
while ( v < 5 )
  v = v + 1;
```

The initial value of 0 must be loaded into the register reg_v before entering the loop. We make this loading explicit as follows.

```
load v = 0;
while ( v < 5 )
  v = v + 1;
```

- *At the completion of a data-dependent loop body.* If v is modified in the body of the loop, then the final value it assumes must be loaded into the register reg_v before the next iteration of the loop. Returning to the previous example, the resulting code segment becomes

```
load v = 0;
while ( v < 5 )
  load v = v + 1;
```

Note that if originally the HardwareC description is as above, with the loads explicitly specified, then since all internal states have been updated explicitly, no loads need to be introduced.

- *At the end of its defining block.* We stated earlier that a static variable needs to retain its last assigned value for the next invocation of the model. If the final value of v is different from its register value, then we need to load this final value into the register reg_v. Consider the defining block of v below.

```
{  static v[4];              /* declaring v */
   v = new_data;
}
```

The final value `new_data` needs to be loaded into reg_v before exiting the statement block. Specifically,

```
{  static v[4];             /* declaring v */
   load v = new_data;
}
```

Note that if we originally started out with the above description, then the explicit load `load v = new_data` resets the final value of v. Therefore, no additional load is necessary.

Execution delay computation

After making all the register loading explicit, we can now describe the basic timing semantics of HardwareC. We do not consider timing and resource constraints at this stage; they are incorporated in the form of constraint graphs and are described in Section 4.2. The rules described in this section serve only to give a lower-bound on the number of clock cycles that are required to execute a particular statement (or group of statements) in HardwareC.

The execution delay for an explicit load assignment is one cycle. Explicit I/O reads and writes to global ports take one cycle each. Message passing sends and receives take at least one cycle, possibly more, to establish handshake. Arithmetic, relational, and logic expressions are assumed to take zero time. Operation chaining via *combinational coalescing* (Section 3.3.2) is applied to cluster together combinational logic blocks, which can be passed to logic synthesis for delay and area estimates. Therefore, the zero-delay assumption serves as a lower bound.

The execution delay for a sequential compound statement $\mathcal{C}_s$ consisting of $\{s_1, s_2, \ldots, s_k\}$ statements is equal to the sum of the execution delays of the individual statements, e.g.,

$$execdelay(\mathcal{C}_s) = \sum_{i=1}^{k} execdelay(s_i)$$

The execution delay for a parallel compound statement $\mathcal{C}_p$ is equal to the maximum of the execution delays of the individual statements, e.g.,

$$execdelay(\mathcal{C}_p) = \max_{i=1}^{k} execdelay(s_i)$$

Finally, the execution delay for a data-parallel compound statement $\mathcal{C}_d$ is equal to the sum of the execution delays of the critical path, e.g.,

$$execdelay(\mathcal{C}_d) = \sum_{j} execdelay(s_j)$$

where s_j lies on the critical path of statements. Note that $execdelay(\mathcal{C}_d) \geq execdelay(\mathcal{C}_p)$.

The execution delay for a procedure (or function) call is equal to the execution delay of the called procedure. The execution delay for a conditional is equal to the execution delay of the selected branch. The execution delay of a loop is equal to the number of iterations times the execution delay of the loop body.

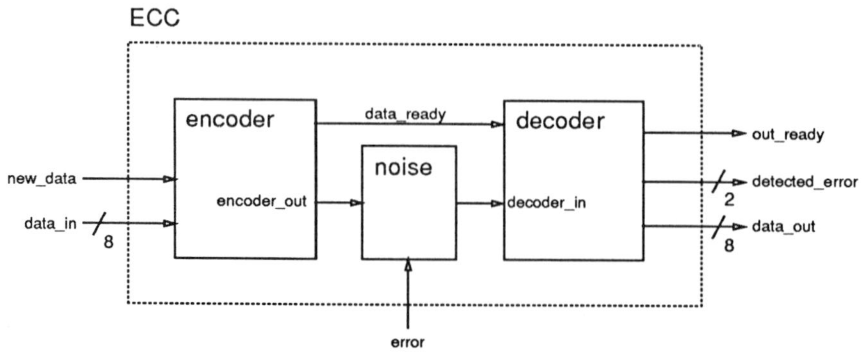

Figure 2.8: Block diagram of the error-correction code design.

2.1.5 A HardwareC example

A complete HardwareC example is given in this section to illustrate the features of the language. The target design is an *error correction system* that models the transmission of digital data through a noisy serial line. Data is read in parallel at the input, encoded with parity information, sent along a serial line (where transmission errors can be inserted), then decoded (possibly correcting single-bit transmission errors), and finally written out in parallel. A block diagram of the example is shown in Figure 2.8.

The HardwareC description for the topmost block, called *ecc*, consists of three concurrent processes: `encoder`, `decoder`, and `noise`. The block takes as input the signals: `new_data`, `data_in`, and `error`. The rising edge of `new_data` indicates that the 8-bit bus `data_in` is ready to be sampled. Noise on the transmission line is modeled by the `error` signal. The output-s for *ecc* are `data_out`, `detected_error`, and `out_ready`. The rising edge of `out_ready` indicates that the recovered data is available on the output bus `data_out`. Status of the error correction is described in the 2-bit signal `detected_error`: "00" if no errors were detected, "01" if single bit error was detected and corrected, and "11" if multiple bit errors were detected. The *ecc* block is described below.

```
block ecc(new_data, data_in, error, data_out,
      detected_error, out_ready)
  in  port   new_data;          /* start signal  */
  in  port   data_in[8];        /* input data    */
  in  port   error;             /* noise signal  */
  out port   data_out[8];       /* output data   */
  out port   detected_error[2]; /* error flag    */
```

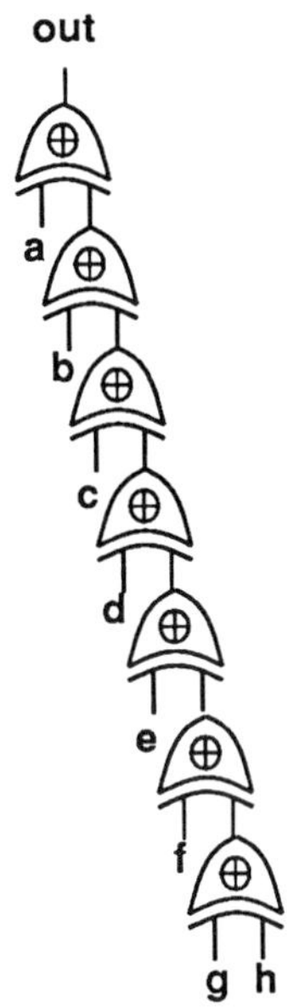

Figure 2.9: Comparison between (a) linear chain and (b) parity tree versions of the *PARITY* template for n equal to 8.

```
  out port    out_ready;            /* data out ready */
<
  boolean encoder_out, decoder_in, data_ready;

  encoder(data_in,new_data,encoder_out,data_ready);
  noise(error,encoder_out,decoder_in);
  decoder(decoder_in,data_ready,data_out,detected_error,out_ready);
>
```

The *encoder* process waits until the rising edge of the ready signal `new_data`, whereupon it samples `data_in`. After computing the parity bits, the process serially transmits the encoded data (8 bits of data and 8 bits of parity information) on the serial line `encoder_out`. An output ready signal `data_ready` is asserted indicating the first bit of the 16-bit data frame. The data frame is organized in the following order:

$$[D0, \ldots, D7, X, R0, R1, R2, C0, C1, C2, T]$$

where $D0, \ldots, D7$ is the data to be transmitted and X is any constant value (0 or 1). The other transmitted signals are defined as follows.

$$\begin{aligned} R0 &= D0\ xor\ D1\ xor\ D2 \\ R1 &= D3\ xor\ D4\ xor\ D5 \\ R2 &= D6\ xor\ D7\ xor\ X \\ C0 &= D0\ xor\ D3\ xor\ D6 \\ C1 &= D1\ xor\ D4\ xor\ D7 \\ C2 &= D2\ xor\ D5\ xor\ X \\ T &= D0\ xor\ \ldots xor\ D7\ xor \\ & \quad R0\ xor\ R1\ xor\ R2\ xor \\ & \quad C0\ xor\ C1\ xor\ C2\ xor\ X \end{aligned}$$

To compute the parity, a template function *PARITY* is invoked by the encoder process. *PARITY* is simply an n-bit XOR function that illustrates the use of function templates in the description. A straightforward way to implement the n-bit XOR is by a linear chain, shown in Figure 2.9(a) and described below.

```
/*
 *  Performs exclusive OR's of all the bits in the operand
 *     - linear chain version
 */
template function PARITY(indata) with (size)  return boolean[1]
  in boolean indata[size];
{
  int       i;
  boolean   temp;

  temp = 0;
  for i = 0 to size-1 do
    temp = temp xor indata[i];
  return_value = temp;
}
```

To improve the performance, however, the parity can be decomposed into groups of 4 signals, where each group is implemented by a parity tree. This is shown in Figure 2.9(b) for `n` equal to 8. The revised template description controls the height of the parity tree based on the value of `size`.

```
/*
 *  Performs exclusive OR's of all the bits in the operand
 *     - size 4 parity tree version
 */
template function PARITY(indata) with (size)  return boolean[1]
  in boolean indata[size];
{
  int       i, offset;
  boolean   temp;

  temp = 0;
  if ( size < 4 ) {
    for i = 0 to size-1 do
      temp = temp xor indata[i];
```

```
  }
  else if ( size < 16 ) {
    /* groups of 4 */
    for i = 0 to (size/4-1) do
      temp = temp xor (indata[4*i]   xor indata[4*i+1]) xor
                      (indata[4*i+2] xor indata[4*i+3]);
    for i = 4*(size/4) to size-1 do
      temp = temp xor indata[i];
  }
  else {
    /* groups of 8 */
    for i = 0 to (size/8-1) do {
      temp = temp xor
                 ( (indata[8*i]   xor indata[8*i+1])  xor
                   (indata[8*i+2] xor indata[8*i+3]) ) xor
                 ( (indata[8*i+4] xor indata[8*i+5])  xor
                   (indata[8*i+6] xor indata[8*i+7]) );
    /* remaining groups of 4 */
    offset = 8*(size/8);
    for i = 0 to ((size-offset)/4-1) do {
      temp = temp xor (indata[4*i+offset]   xor indata[4*i+offset+1]) xor
                      (indata[4*i+offset+2] xor indata[4*i+offset+3]);
    for i = 4*(size/4) to size-1 do
      temp = temp xor indata[i];
  }
  return_value = temp;
}
```

The HardwareC description for the *encoder* process is given next.

```
/*
 *  Encoder process
 */
process encoder(data_in,new_data,encoder_out,data_ready)
  in   port data_in[8];
  in   port new_data;
  out  port encoder_out;
  out  port data_ready;
[
  int      i;
  boolean  output_data[16];

  /* wait for incoming data */
  while ( !new_data) ;

  /* sample parallel input data */
  output_data[0:7] = read (data_in);
  output_data[8] = 0;
  output_data[15] = 0;

  /* calculate parity bits for row and column */
  for i = 0 to 2 do {
    output_data[i+9] = PARITY (
        output_data[3*i] @ output_data[3*i+1] @
        output_data[3*i+2] ) with (3);

    output_data[i+12] = PARITY (
```

```
          output_data[i] @ output_data[i+3] @
          output_data[i+6] ) with (3);

      output_data[15] = PARITY (
        output_data[15]    @ output_data[3*i]    @
        output_data[3*i+1] @ output_data[3*i+2] ) with (4);
    }

    /*  write output stream, the first bit is
     *  indicated by a pulse in data_ready  */
    <
      [
        write data_ready = 1;
        write data_ready = 0;
      ]

      for i = 0 to 15 do {
        write encoder_out = output_data[i:i];
      ]
    >
  ]
```

The *noise* process is modeled by an XOR between the serial data stream and an error signal. The serial data is "flipped" whenever the `error` signal is asserted. The process is described below.

```
/*
 *  Noise generator on the serial data stream
 */
process noise(error,encoder_out,decoder_in)
  in   port error;
  in   port encoder_out;
  out  port decoder_in;
{
  decoder_in = encoder_out xor error;
}
```

The *decoder* process waits until the rising edge of the ready signal `data_ready`, then samples the 16-bit data frame serially. Single-bit error-correction is performed using the parity bits in the serial stream to recover the original 8-bit data. The output signal `err` indicates whether no error ("00"), single-bit error ("01"), or multiple bit error ("11") has occurred.

```
/*
 *  Decoder process
 */
# define NO_ERROR          0b00
# define ONE_ERROR         0b01
# define MULTIPLE_ERRORS   0b11

process decoder(decoder_in, data_ready, data_out, err, out_ready)
  in   port decoder_in;
  in   port data_ready;
```

```
    out  port data_out[8];
    out  port err[2];
    out  port out_ready;
  [
    int       i;
    boolean   input_data[16];
    boolean   row_parity[3];
    boolean   column_parity[3];
    boolean   global_parity;
    boolean   error[2];

    /*  rising edge of data_ready */
    while (!data_ready) ;

    /*  sample input stream  */
    for i = 0 to 15 do [
      input_data[i] = read (decoder_in);
    ]

    /*  compute parity check on the input data  */
    global_parity = input_data[15];

    for i = 0 to 2 do {
      row_parity[i] = PARITY(
        input_data[3*i]   @ input_data[3*i+1] @
        input_data[3*i+2] @ input_data[i+9] ) with (4);

      column_parity[i] = PARITY(
        input_data[i]   @ input_data[i+3] @
        input_data[i+6] @ input_data[i+12] ) with (4);

      global_parity = PARITY(
        global_parity     @ input_data[3*i]   @
        input_data[3*i+1] @ input_data[3*i+2] ) with (4);
    }

    /*  error correction  */
    if (!global_parity) {
      if ((row_parity[0:2] == 0b000)&(column_parity[0:2] == 0b000))
          error = NO_ERROR;
      else
          error = MULTIPLE_ERRORS;
    }
    else {  /* Single error */
      if (row_parity[0:0])
      {
        if (column_parity[0:0])
            input_data[0:0] = !input_data[0:0];
        else if (column_parity[1:1])
            input_data[1:1] = !input_data[1:1];
        else if (column_parity[2:2])
            input_data[2:2] = !input_data[2:2];
      }
      else if ( row_parity[1:1])
      {
        if (column_parity[0:0])
            input_data[3:3] = !input_data[3:3];
```

```
        else if (column_parity[1:1])
            input_data[4:4] = !input_data[4:4];
        else if (column_parity[2:2])
            input_data[5:5] = !input_data[5:5];
      }
      else if ( row_parity[2:2])
      {
        if (column_parity[0:0])
            input_data[6:6] = !input_data[6:6];
        else if (column_parity[1:1])
            input_data[7:7] = !input_data[7:7];
      }
      error = ONE_ERROR ;
    }

    /*  write parallel data and error flags  */
    <
      write data_out = input_data[0:7];
      write err = error;
      [
        write out_ready = 1;
        write out_ready = 0;
      ]
    >
  ]
```

2.2 Overview of the Synthesis Flow

The algorithms developed in this research have been implemented in two programs, *Hercules* and *Hebe*. Hercules performs the front-end parsing and behavioral optimizations to identify the parallelism in the input description. Hebe maps the behavior to structure by performing scheduling and allocation, generating a synchronous logic-level implementation that realizes the input behavior. This section describes the overall synthesis flow.

To decouple the system from a particular input language, synthesis transformations at the behavioral, structural, and logic levels are defined and applied on the following internal models:

- *Behavioral Intermediate Form* (BIF): An *abstract syntax tree* is used as the internal model at the behavioral level to capture imperative semantic languages.
- *Sequencing Intermediate Form* (SIF): A hierarchical *sequencing graph* is used to capture both control-flow and data-flow at the structural synthesis level.
- *Structure/Logic Intermediate Form* (SLIF): At the logic level, the internal

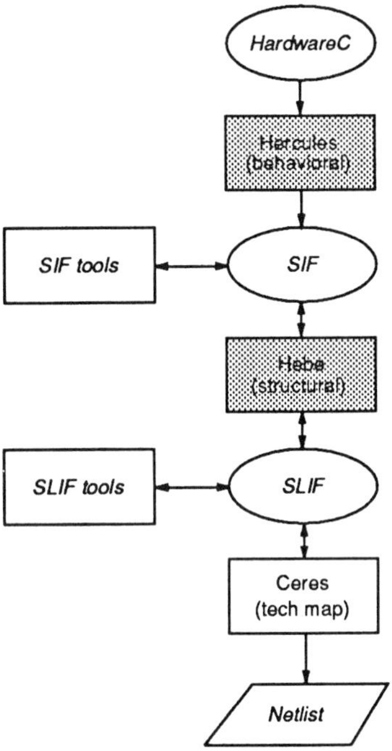

Figure 2.10: The Stanford Olympus Synthesis system.

model is based on a synchronous logic network representation that supports hierarchy and both mapped and unmapped Boolean expressions. The SLIF is an interchange format used by other tools operating at the logic level.

The system can be seen as a set of synthesis tools operating on, and bridging across, the internal models. Hercules transforms a HardwareC description into a set of SIF models and Hebe transforms a SIF model into one or more SLIF implementations. Accurate delay and area information extracted from logic synthesis are fed back to guide the scheduling and allocation tasks in Hebe. Using internal models means different input languages can be supported simply by parsing the language to the appropriate internal model.

Hercules and Hebe form the front-end to the *Stanford Olympus Synthesis system*, a research project in computer-aided synthesis at Stanford University [MKMT90]. A block diagram of the system is shown in Figure 2.10. *Ceres* [MM90] performs technology mapping to transform a logic network into elements of a given technology library, e.g. LSI, Actel, or other cells from a semi-custom library. Other tasks, such as partitioning, simulation, and logic minimization, operate at either the SIF or SLIF levels.

Synthesis can be fully automated to transform an input HardwareC description directly to a logic-level implementation. Hercules and Hebe also support user-driven synthesis, where a designer can intervene and drive high-level decisions based on an evaluation of the possible design tradeoffs.

2.2.1 Hercules: Behavioral synthesis

Hercules identifies the parallelism in the HardwareC description by performing behavioral transformations. These transformations are categorized as either *user-driven* or *automatic*. User-driven transformations modify the calling hierarchy of the description by either expanding procedure calls in-line or mapping operators to library models. They are controlled by the user and are optionally performed. In contrast, automatic transformations are always performed to optimize the behavior for a given calling hierarchy. These transformations include compiler optimizations such as dead-code elimination, constant and variable propagation, loop unrolling, and common subexpression elimination. Logic operations in the description are clustered to form blocks of combinational logic that are passed directly to logic synthesis for minimization and delay/area estimates.

Behavioral transformations operate on the BIF model, which is generated by parsing the input HardwareC description. Upon completing the behavioral transformations, the optimized BIF is translated to the sequencing graph representation described in SIF.

2.2.2 Hebe: Structural synthesis

Hebe takes as input a hardware behavior represented by a sequencing (SIF) graph, and produces a synchronous logic-level implementation that realizes the original behavior. Hebe allows the designer to experiment with different design goals relating to area and/or performance of the final design. For example, the design goal may be to find the *minimum area* configuration that satisfies the timing constraints.

The input to Hebe consists of a sequencing graph model and the following constraints:

- *Timing constraints* – that specify upper and lower bounds on the time separation between activation of operations.

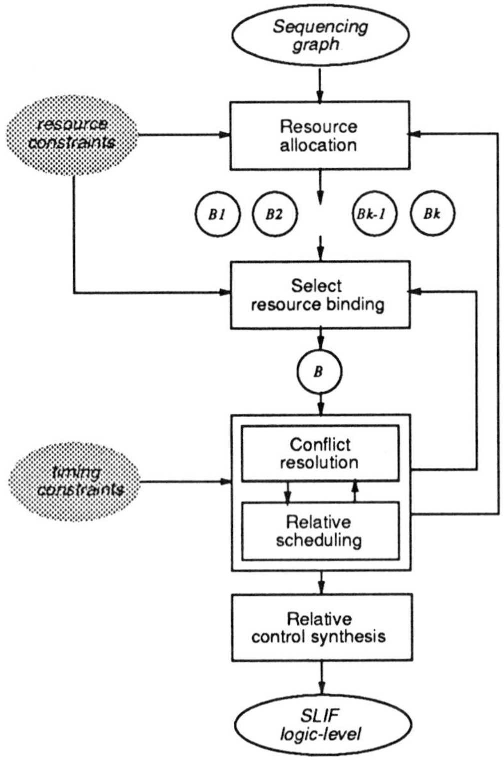

Figure 2.11: Structural synthesis flow in Hebe.

- *Resource constraints* – that both limit the number of instances allocated for each resource type and partially bind operations to specific allocated resources.
- *Cycle time* – for the final synchronous logic implementation.

These constraints can be specified either in the input description or entered interactively by the designer. Note that they are not mandatory. For example, if the cycle time is not given, then the cycle time is by default equal to the critical combinational logic delay in the final logic-level implementation. The final implementation contains both data-path and control. The data-path is an interconnection of functional units, registers and multiplexers.

Hebe performs resource allocation and binding before scheduling. This strategy has the advantages of providing scheduling with detailed interconnection delays and incorporating partial binding information to limit the number of design choices. The structural synthesis flow in Hebe is illustrated in Figure 2.11.

The major synthesis steps are summarized below.

- *Resource allocation.* The designer first allocates the resources (i.e. functional units) subject to resource constraints. Allocation is guided by the design goal, i.e. minimum area or minimum latency or a combination of both.

- *Resource binding.* Given a resource allocation, many possible bindings of operations to specific resources exist. These resource bindings correspond to different design choices and represent the *design space* for the design. They are denoted by $B_1, \ldots, B_k$ in Figure 2.11. A particular resource binding is selected as a candidate for synthesis. The search strategy is driven by a combination of area, interconnect, and delay costs.

- *Resource conflict resolution.* A resource binding implies a certain configuration of hardware sharing. In general, resource conflicts can arise when a resource is accessed simultaneously by multiple operations. These conflicts can be resolved by serializing operations bound to the same resource that can otherwise execute in parallel. Timing constraints must still be satisfied after conflict resolution.

- *Relative scheduling.* Once resource conflicts have been resolved, operations are assigned to control steps subject to synchronization and timing requirements. The formulation is based on relative scheduling.

- *Data-path and control generation.* Generation of a data-path involves instantiating the allocated functional units, performing register folding to minimize the number of registers, and constructing the interconnect. Control is synthesized for a relative schedule as an interconnection of finite-state machines using relative control synthesis.

Conflict resolution, relative scheduling, relative control synthesis and optimization are formulated on a *constraint graph model* that is derived from the sequencing graph model under detailed timing constraints. Descriptions and analyses of these formulations are presented in subsequent chapters.

Chapter 3

Behavioral Transformations

This chapter describes the transformations that are performed at the behavioral level in Hercules. Behavioral transformations identify the inherent inter-operation parallelism in the behavior. Understanding the parallelism can give an indication of the *fastest* design synthesis can produce, assuming in the final implementation each operation is implemented by a dedicated hardware component. While this assumption may not be realistic due to area and interconnection limits, computing the performance under this assumption is important as a bound on the given behavior.

The input HardwareC description is parsed by Hercules and translated into an abstract syntax tree representation called the Behavioral Intermediate Form (BIF). BIF is the underlying representation for all behavioral transformations. User-driven transformations are first performed by the designer to determine the hierarchy of the BIF induced by the procedure call mechanism in HardwareC, called its *calling hierarchy*. The calling hierarchy affects the granularity of hardware sharing in subsequent synthesis steps because each model in the behavior corresponds to a hardware resource that can be shared and allocated. If the original calling hierarchy is acceptable, then this step is optional. Next, the BIF is optimized by automatic behavioral transformations that include techniques used by optimizing compilers. The resulting optimized BIF is translated to the sequencing graph model described in SIF.

This chapter is organized as follows. Section 3.1 describes the BIF model. Sections 3.2 and 3.3 present the user-driven and automatic behavioral transformations, respectively. Finally, Section 3.4 summarizes the behavioral transformations applied in Hercules.

Name	*Description*
Nop	No-op
Block	Block of statements
Call	Procedure or function call
Loop	Data-dependent loop
For-loop	Fixed-bound loop
Switch	Multi-way conditional
Op	Unary or binary operator
Assign	Assignment to variable
Ipc	Message passing send/receive
Io	I/O read and write

Table 3.1: Operator nodes in the Behavioral Intermediate Form.

3.1 Behavioral Intermediate Form

An *abstract syntax tree*, similar to that used in compiler designs [ASU87], is adopted as the underlying representation for behavioral transformations. An abstract syntax tree is constructed from a set of nodes, where each node represents an operator in the language and its children represent the operands. We use the terms syntax tree, abstract syntax tree, and BIF interchangeably throughout this chapter.

There are two types of BIF nodes: *value nodes* and *operator nodes*. Value nodes represent constants and variables in the language. They form the leaves of the syntax tree. Operator nodes, on the other hand, represent operations in the language. They can be either leaves or interior nodes of the syntax tree depending on the type of operation being represented. Table 3.1 lists the set of operator node types which is sufficient to model HardwareC.

For example, an addition is represented by an *Op* node. The operands of the addition correspond to the children of the *Op* node. Likewise, a compound statement block is represented by a *Block* node. Each statement in the compound statement corresponds to a child of the *Block* node. The two examples are illustrated in Figure 3.1. Squares in the figure denote value nodes and circles denote operator nodes.

User-defined variables are *scoped*, which means that a variable is valid only in its defining block. In HardwareC, variables are declared in statement blocks; therefore, variables declared in a *Block* node are valid only in its children. The

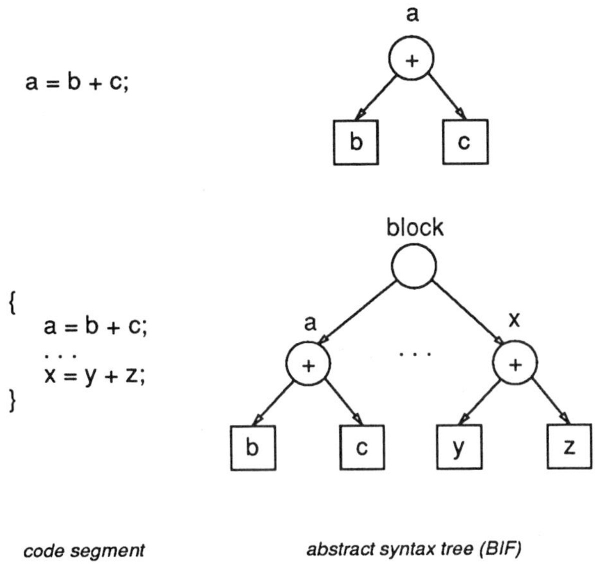

Figure 3.1: Examples of the translation to the Behavioral Intermediate Form.

scoped nature of BIF is important because it allows simplification of transformations involving duplication of statements, such as in-line expansion of procedures and unrolling of loops. By basing all transformations on the BIF, Hercules can be organized as a multiple pass optimization system where each optimization constitutes a traversal of the BIF.

3.2 User-driven Behavioral Transformations

The calling hierarchy of the input description can be modified by the designer using user-driven behavioral transformations. These transformations include *in-line expansion of model calls* and *mapping of operators to models.*

Selective in-line expansion of model calls.

A call to a model can be replaced by the functionality of the called model. In-line expansion increases the scope of behavioral transformations because optimizations can now be applied across the original calling hierarchy. This is accomplished on the BIF by replacing a *Call* node with the BIF of the called model.

Consider for example a call to the `Adder` procedure implementing the addition of two numbers. Assume the call is used to increment a variable v by one, so that one operand of the addition is v and the other is the constant 1. By expanding the call to `Adder`, the addition logic can be optimized taking into account that only an increment is needed instead of a full addition. Without in-line expansion such context-based optimization is not possible.

However, always performing in-line expansion may lead to excessively large hardware implementations. The reason is because the hardware components used to implement the expanded model are now *dedicated* to that particular call. These hardware components cannot be shared with other calls in the description. The decision of which model calls to expand is left to the discretion of the designer because it involves the subjective choice over the granularity of hardware sharing.

Selective operator to library mapping.

Operators in HardwareC, such as "+" or "−", can be mapped into calls to library template models. The motivation for operator mapping is to give the designer ability to select a specific implementation for an operator. Operator mapping is accomplished on the BIF by replacing an *Op* node with a *Call* node to the appropriate template model.

For the sake of illustration, consider the following code segment with three additions. The first addition is with 4-bit values (i.e. x and a are 4-bit variables) and the second and third additions are with 8-bit values.

```
boolean    x[4], a[4];
boolean    y[8], b[8], c[8];
boolean    z[8], d[8], e[8];

x = a + 1;        /* 4-bits */
y = b + c;        /* 8-bits */
z = d + e;        /* 8-bits */
```

If the first two additions are mapped to the `ripple_adder` template of Figure 2.5 and the last addition is mapped to a `fast_adder` template, then the above code segment is transformed to the following:

```
x = ripple_adder(a, 1, 0) with (4);
y = ripple_adder(b, c, 0) with (8);
z = fast_adder(d, e, 0) with (8);
```

The designer can use operator mapping to specify the "types" of hardware functional units that will be used to implement a design. If no mapping is given, then by default operators are implemented as combinational logic. Because of the size of combinational multipliers and dividers, the multiply (*) and divide (/) operators are always mapped to library templates prior to structural synthesis in Hebe.

3.3 Automatic Behavioral Transformations

Automatic behavioral transformations are carried out without human intervention. They optimize the behavior using techniques similar to those found in optimizing compilers [ASU87]. Compiler-like transformations are used in some high-level synthesis systems [WT89, Tri87, Sno78]. Hercules performs the following transformations, described below in the order of application.

- `For`*-loop unrolling.* Fixed-iteration loops are unrolled to increase the scope of subsequent optimizations. This is accomplished in BIF by replacing the *For-loop* node with a *Block* node, where each child of the block node is a duplicate of the loop body for a particular value of loop index. The example below illustrates the loop unrolling transformation.

  Before:
  ```
  for i = 0 to 3 do
          result[i] = a[i] & b[i];
  ```

  After:
  ```
  result[0] = a[0] & b[0];
  result[1] = a[1] & b[1];
  result[2] = a[2] & b[2];
  result[3] = a[3] & b[3];
  ```

- *Meta-variable resolution.* All integer variables and expressions are resolved and replaced by the evaluated values. Only Boolean variables remain in the syntax tree after this optimization.

- *Constant conditional elimination.* A conditional statement where the conditional variable is a constant value is replaced by the selected branch.

- *Constant and variable propagation.* Constants and variables are propagated as far as they can, replacing a variable reference by its last assigned value. Expressions with constant operands are evaluated and reduced during this propagation.

- *Reference stack resolution.* Multiple and conditional assignments to variables are resolved by creating *multiplexed values.* Section 3.3.1 describes the details of the reference stack algorithm.

- *Common sub-expression elimination.* Redundant operations that produce the same result, i.e. identical operands and operator, are removed.

- *Dead-code elimination.* Dead-code refers to operations that do not affect the output of the model (i.e. ports of the model). They can be eliminated without affecting the externally visible I/O behavior of the model. Dead-code elimination is recursively backtracked to remove all unnecessary operations.

- *Combinational conditional collapsing.* A conditional whose branches contain only combinational logic is collapsed to form a single statement block containing combinational logic. This transformation allows logic synthesis to be applied to a larger scope of operations. An example is described below.

```
Before:         if ( cond ) {
                  a = b & c;
                  d = e | f;
                } else {
                  h = i ^ j;
                  x = y ^ z;
                }

After:          {
                  a = b & c;
                  d = e | f;
                  h = i ^ j;
                  x = y ^ z;
                }
```

  Note that in the current implementation, *don't care* information resulting from mutual-exclusiveness of the branches is not preserved after collapsing.

- *Combinational logic coalescing.* Combinational logic operations in the BIF are clustered to form *expression blocks.* Each expression block represents the largest scope without crossing control step boundaries in which combinational logic synthesis can be applied. Details of the coalescing algorithm are described in Section 3.3.2.

Many of the above transformations are well-documented in texts on compiler design. The reference stack and combinational logic coalescing techniques are new. They are described in the next two sections.

3.3.1 Reference stack

The *reference stack algorithm* efficiently resolves constant and variable propagation, conditional assignment to variables, and multiple assignment to variables. By using the reference stack, many of the control steps that are traditionally needed to implement assignment to variables can be eliminated. This leads to an improvement in the latency of the resulting hardware.

When an imperative-semantic program references a particular variable at different locations in the code, it may reference different *values* of that variable, depending on whether assignments were made between the references. The *value* of a variable in a program is defined to be the most recently assigned value of that variable. The algorithm associates a *stack* with each variable in the program. The top of stack represents the most recent value at a given stage of translation.

For the sake of illustration, consider the following code segment. The first column shows the HardwareC code. Temporary variables are introduced by the parser to hold the results of any binary or unary operation, denoted by $T1$ and $T2$ in the example below. The second column is the first column augmented with temporary variables. The third column shows the code segment after variable references are replaced by the top of reference stack. The underlined entries highlight the references that have been changed.

$$
\begin{array}{lclcl}
a = b + c; & & T1 = b + c; & & T1 = b + c; \\
 & & a = T1; & & a = T1; \\
d = a; & \Longrightarrow & d = a; & \Longrightarrow & d = \underline{T1}; \\
a = a + 1; & & T2 = a + 1; & & T2 = \underline{T1} + 1; \\
 & & a = T2; & & a = T2; \\
e = a; & & e = a; & & e = \underline{T2};
\end{array}
$$

Whenever an assignment is made to a variable, subsequent references to that variable results in referencing the *value* and not the *variable* itself. In its simplest form, the reference stack algorithm performs constant and variable propagation.

Variables in HardwareC are categorized into local variables (which are in addition categorized into **static**, **boolean** and **int** variables) and external ports. Whereas local variables are not externally visible, ports connect external signals with computations within the model. Only explicit *write* commands can change the external value of a given output port. Once written, a port retains its last written value until it is rewritten. Therefore, only local variables are candidates for the reference stack algorithm.

Treatment of Conditionals

The situation is more complicated in the case of conditional assignments. Consider for example the following code segment.

$a = T1;$
if $(condition)$
    $a = T2;$
$x = a;$

Variable a is assigned the value $T2$ if $condition$ is true. The question is "what is the value of x?" In some synthesis approaches, variable a is implemented by a register because of this conditional assignment. Each assignment to a results in loading the register with the assigned value. The above code segment in this case is implemented as follows.

load $a = T1;$
if $(condition)$
    load $a = T2;$
$x = a;$

The **load** operator updates register a with a given value. Since loading of a register requires one control step to execute, the code segment above requires at least two control steps to execute. However, the conditional only *steers* the appropriate inputs to x based on the value of $condition$. Therefore, a simpler approach is to implement x as a multiplexer of $T1$ and $T2$ that depends on the value of $condition$, as shown below.

$X1 = (condition \wedge T2) \vee (\overline{condition} \wedge T1);$
$x = X1;$

HardwareC supports the first approach by explicitly modeling the loading of variables in the language (as shown in the example above). For the case where not all assignments in the description are explicit loads, the second approach is supported automatically when processing the HardwareC description during behavioral transformations using the reference stack algorithm.

We present in the remainder of this section the details of the reference stack algorithm. Let the reference stack for a variable v be denoted by $RS(v)$. $RS(v)$

is initialized according to the variable type of v: if v is a Boolean variable then $RS(v)$ is initialized to zero; if v is a static variable then $RS(v)$ is initialized to the variable v itself.

The algorithm traverses the syntax tree in pre-order. Let the traversed node be denoted as $node$. If $node$ is a value node, then it is replaced by the most recent value of $node$: $node \leftarrow RS(node)$. If $node$ is an assignment to a variable v, then $RS(v)$ is updated with the assigned value. If $node$ is a conditional with a conditional variable c having size $|c|$, then the algorithm performs the following steps.

Let $\mathcal{M}(node)$ denote modified variables in any conditional branch. For each variable $v \in \mathcal{M}(node)$, a new entry is pushed onto the reference stack $RS(v)$ *for each different case* of the conditional. For example, if c is an 1-bit Boolean variable ($|c| = 1$), then two new entries are added to the top of the stack; one corresponds to the case $c = 0$ and the other corresponds to the case $c = 1$. Likewise, if c is a 3-bit Boolean variable ($|c| = 3$), then eight (2^3) new entries are added; each entry corresponds to one of the eight possible values of c. New stack entries are initialized to the previous top of stack value.

The algorithm is recursively applied to each of the conditional branches. For a branch j, the top of the reference stack refers to the newly created stack entry corresponding to the case $(c = j)$. Assignments and references in a branch access and modify the appropriate stack entry only. Once all the branches have been traversed, the algorithm pops off these $2^{|c|}$ stack entries and examines their values.

The following two cases are possible, where $\{RS_0(v), RS_1(v), \ldots, RS_{2^{|c|}-1}(v)\}$ denote the $2^{|c|}$ stack entries for $v \in \mathcal{M}(node)$.

- *All entries are identical*: The new top of stack is updated with the value $RS_0(v) = \cdots = RS_{2^{|c|}-1}(v)$.
- *All entries are not identical*: A *multiplexed value* X is created, defined as $X = \sum_{i=0}^{2^{|c|}-1} (c = i) \cdot RS_i(v)$. The new top of stack is updated with the value X. Subsequent references to v will refer to X.

An example of the algorithm is shown in Figure 3.2. The first and second columns represent the code segment before and after application of the algorithm, respectively. The reference stack for variable a is shown, which is updated whenever an assignment to a occurs. In the case of the conditional on $cond$, two entries are pushed onto the top of stack. These stack entries are initialized

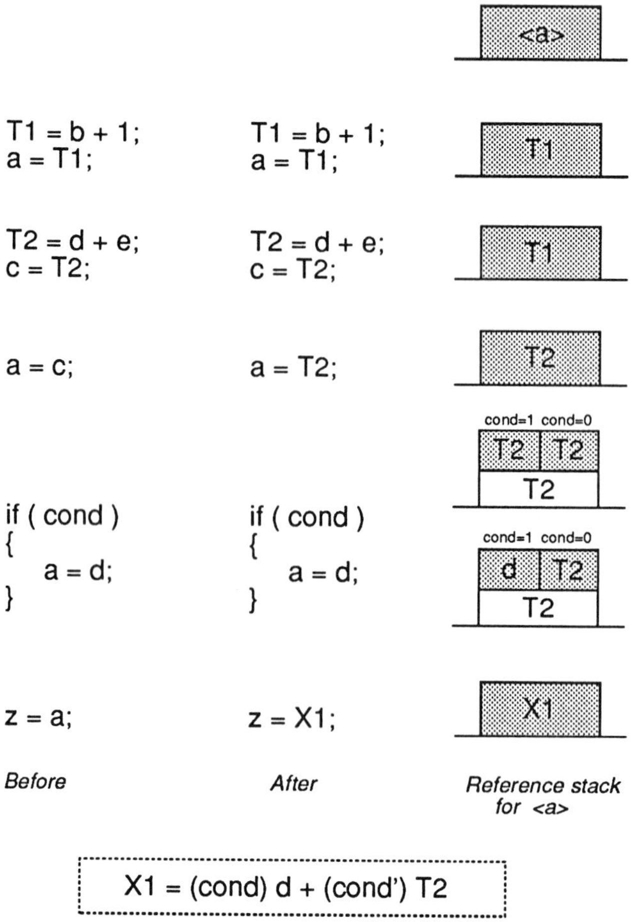

Figure 3.2: Example of the reference stack algorithm applied to a code segment. The reference stack for variable a is shown.

to the previous stack value $T2$. Since in the $cond = 1$ branch an assignment to a was made, a multiplexed value $X1$ is created upon exiting the conditional.

Treatment of Loops

We now consider the case when the traversed node is a loop with a single-bit condition variable $exit$, i.e. $|exit| = 1$. Let $\mathcal{M}(node)$ denote modified variables in the loop body. Note that all loops with fixed iteration bounds have already been unrolled. Therefore, $node$ is a data-dependent loop such as a while loop.

Similar to the case for conditionals, two new entries are pushed onto the top of stack for each variable $v \in \mathcal{M}(node)$. One entry corresponds to the case when the loop is not entered ($exit = 1$) and is initialized to the previous value of $RS(v)$. The other entry corresponds to the case when the loop is entered ($exit = 0$) and is initialized to the variable v itself.

The loop body is traversed with the ($exit = 0$) entry as the top of stack. Upon completion, the two entries for each variable are examined. If the entries are identical, then they are popped off and the new top of stack is updated to this value. Otherwise, the variable has been modified in the loop, in which case it is stored into a register, or equivalently, it is classified as a **static** variable. The register first loads the value of the ($exit = 1$) entry before entering the loop. Each time the loop is repeated, it loads the value of the ($exit = 0$) entry. For this case, the new top of stack is updated with the variable itself.

Consider for example the following code segment.

```
a = T1;
while (val = 1) {
        reference a;
        a = T2;
}
x = a;
```

Variable a is first assigned to the value $T1$, then it is assigned the value $T2$ in the while loop. After the algorithm, variable a is made into a register because it is referenced before written in the loop. The final code segment is given below.

```
static a;

load a = T1;
while (val = 1) {
        reference a;
        load a = T2;
```

```
}
x = a;
```

The declaration **static** indicates that a is implemented as a register. Loading of register $a = T1$ can be removed if the register already contains the value $T1$ before entering the loop. Loading of register $a = T2$ can take place in parallel with the restarting of the while loop. Therefore, it does not increase the overall latency of the design at hardware execution time.

Advantages and disadvantages of the Reference Stack

This approach has three important advantages. First, in many cases it avoids requiring additional control steps to implement conditional assignments to variables. This reduces the latency of the resulting hardware. Second, it makes explicit much of the steering logic that is implied by the description. When combined with the combinational logic coalescing step of Section 3.3.2, logic synthesis can optimize not only the data path logic expressions, but also a significant portion of the steering logic. This approach greatly simplifies the generation of interconnect during structural synthesis. Finally, variables that are required to be implemented as registers due to the semantics of the description, i.e. referenced before assigned in a loop, can be directly identified.

A disadvantage of this approach is that the complexity of the algorithm is related to the number of branches and nesting depth of the conditionals. Therefore, the hardware modeling style has a large impact on the execution time of the algorithm.

3.3.2 Combinational logic coalescing

Combinational logic coalescing clusters together combinational logic operations in the description into a number of *expression blocks*. These operations include Boolean AND, OR, XOR, and NOT, relational operators such as greater-than or equality comparisons, and multiplexed values resulting from the reference stack algorithm. If the entire description consists of only combinational logic operations, then a single expression block is created.

An expression block defines the largest scope (without crossing control step boundaries) in which combinational logic synthesis can be applied. There are two advantages in performing coalescing:

- *Link to logic synthesis:* Each expression block is directly passed to logic synthesis for minimization and technology mapping. The results are fed-back as estimates on area and critical path delay that are then used to refine the design and guide subsequent scheduling and allocation.

- *Support for operation chaining:* Often, operation delays are small and many operations can be executed within one clock cycle. Packing multiple operations within a single clock cycle improves the latency of the final hardware. The sum of propagation delays of the chained operations must not exceed the cycle time. Operation chaining is particularly important for ASIC designs because of the extensive use of logic expressions in the hardware specification and is supported by combinational logic coalescing.

An example of coalescing is given below. Originally, the specification consists of a set of Boolean operations and two I/O statements, grouped in a data-parallel statement block, i.e. operations can execute in parallel subject to data-dependencies.

```
{
    T1 = a & b;
    T2 = T1 & c;
    write bus = T1;
    x = read( bus );
    T3 = x | y;
    T4 = z & w;
    T5 = T3 & T4;
}
```

Coalescing produces two expression blocks, one before the I/O statements and one after.

```
{
    {   /*  expression block #1   */
        T1 = a & b;
        T2 = T1 & c;
    }
    write bus = T1;
    x = read( bus );

    {   /*  expression block #2   */
        T3 = x | y;
        T4 = z & w;
        T5 = T3 & T4;
    }
}
```

Combinational coalescing is implemented by clustering the BIF based on data-flow dependencies. Logic operations are gathered into a cluster subject to the restriction that no sequential operations exist that both reference the outputs of operations in the cluster and produce outputs that are referenced by operations in the cluster. The algorithm is straightforward and is not described here.

3.4 Summary of Behavioral Transformations

Behavioral transformations identify the parallelism in the HardwareC description using compiler optimization techniques. They also permit the designer to change the procedure calling hierarchy to control the granularity of hardware sharing in subsequent synthesis steps. The BIF is used as the underlying representation for all transformations.

The optimized BIF is translated into a graph-based abstraction called the *Sequencing Intermediate Form* (SIF), which is the basis for the structural synthesis tasks in Hebe. Behavioral synthesis in Hercules results in a *maximally parallel* sequencing graph where the inter-operation dependencies are due to either data-flow or control-flow dependencies that are explicitly specified in the input description. The SIF is then selectively serialized by Hebe during hardware sharing to meet both resource and timing constraints.

Chapter 4

Sequencing Graph and Resource Model

This chapter presents the graph-based model of hardware behavior that is used as the underlying representation for the synthesis algorithms in Hebe. A hierarchical *sequencing graph* is used to model hardware behavior for synthesis. The hardware behavior is assumed to be synchronous and non-pipelined. As with other hardware models, both control-flow and data-flow dependencies are represented by the sequencing graph model. Its main distinction is that it uniformly supports *concurrency*, *synchronization*, and *detailed timing constraints*. Synthesis algorithms operating on this model can guarantee that the resulting design satisfies the imposed constraints, or indicate when no such solution exists. The use of this model decouples the synthesis algorithms from the HardwareC language. In particular, different hardware description languages can be supported by compiling them into this model. The *Sequencing Intermediate Form* is an ASCII representation of the sequencing graph which is used as an interchange format facilitating communication between different tools, i.e. the Vulcan partitioner [GM90] and Hebe both use the SIF format.

Similar models are used in existing high-level synthesis systems. They include DACON in Flamel [Tri87], YIF in YSC [BCM+88], sequence, data-flow, and control graphs in the Caddy system [CR89], Value Trace in CMU's System Architect's Workbench [TDW+90], and CDFG in Elf [GK84].

This chapter is organized as follows. Section 4.1 describes the semantics of the sequencing graph model. Section 4.2 describes the derivation of a constraint graph model from a sequencing graph model under timing constraints.

Section 4.3 describes the model of hardware resources used in synthesis. Section 4.4 summarizes the major features of the sequencing graph model.

4.1 Basic Definitions

The sequencing graph is a concise way of capturing the partial order among a set of operations. An operation can be a data computation, i.e. an arithmetic operation or a set of combinational logic expressions, called a *logic block*. Alternatively, it can model control-flow constructs, i.e. a call to another model, a conditional or a loop. This model captures the precedence relationship among the operations and defines the execution flow in implementing a given behavior. To be more exact, a sequencing graph is defined as follows.

Definition 4.1.1 *A* **sequencing graph** *is a polar, hierarchical, vertex-weighted, directed acyclic graph, denoted by $G_s(V, E_s, \delta)$. The vertices $V = \{v_0, \ldots, v_N\}$ represent operations to be performed, where v_0 and v_N denote the source and sink vertices, respectively, of the polar graph. Directed edges E_s represent sequencing dependencies among the operations. An integer weight $\delta(v_i)$ is associated with each vertex $v_i \in V$ representing its execution delay.*

A polar graph has a single source vertex and a single sink vertex.

Sequencing dependencies can arise due to data-flow dependencies extracted from the behavioral model (i.e. a value must be written before it can be referenced), explicit sequencing that is specified in the input description (i.e. detect the rising edge of a signal before reading a bus), or resource sharing restrictions that are introduced during structural synthesis (i.e. operations sharing the same hardware resource are serialized to avoid resource conflicts). A directed edge $s_{ij} \in E_s$ from vertex v_i to v_j means that v_j can begin executing only after the completion of v_i; v_i is called a *predecessor* of v_j, and v_j is called a *successor* of v_i.

Vertices are classified into different types according to the operations they perform. For example, the types listed in Table 4.1 are sufficient to model the HardwareC language. Vertices are further categorized as either *simple* or *complex*:

- *Simple:* Simple vertices are primitive computations that do not involve other operations, i.e. arithmetic or logic operations and message passing commands.

Category	*Vertex type*	*Operation represented*
simple	*no-op*	No operation
	computation	Arithmetic, Boolean, or relational expressions
	logic block	Set of combinational logic exprs
	load register	Load internal register
	input/output	I/O port access
	send/receive	Interprocess message passing
complex	*call*	Invoke a procedure
	conditional	Select among several branches
	loop	Fixed or data-dependent iterations

Table 4.1: Vertex types in the sequencing graph model representing HardwareC constructs.

- *Complex:* Complex vertices allow groups of operations to be performed. They include *model calls*, *conditionals*, and *loops*, and are analogous to structured control-flow constructs in most programming and hardware description languages.

The semantic interpretation of the sequencing graph is as follows. A vertex *executes* by performing its corresponding operation. For example, to execute a conditional vertex, operations in the selected branch are executed. Executing a sequencing graph is equivalent to executing the vertices according to the precedence relations implied by the graph starting from the source vertex. A vertex can execute only when its predecessors have completed execution. Since a vertex can have multiple predecessors and multiple successors, the model supports *multiple threads of concurrent execution flow.*

Complex vertices induce a hierarchical relationship among the graphs. A call vertex invokes the sequencing graph corresponding to the called model. A conditional vertex selects among a number of branches, each of which is modeled by a sequencing graph. A loop vertex iterates over the body of the loop until its exit condition is satisfied, where the body of the loop is also a sequencing graph. The sequencing graph is *acyclic* because only structured control-flow constructs are assumed (i.e., no `goto`'s) and loops are broken through the use of hierarchy. All forms of conditional branching are represented as complex vertices in the graph model.

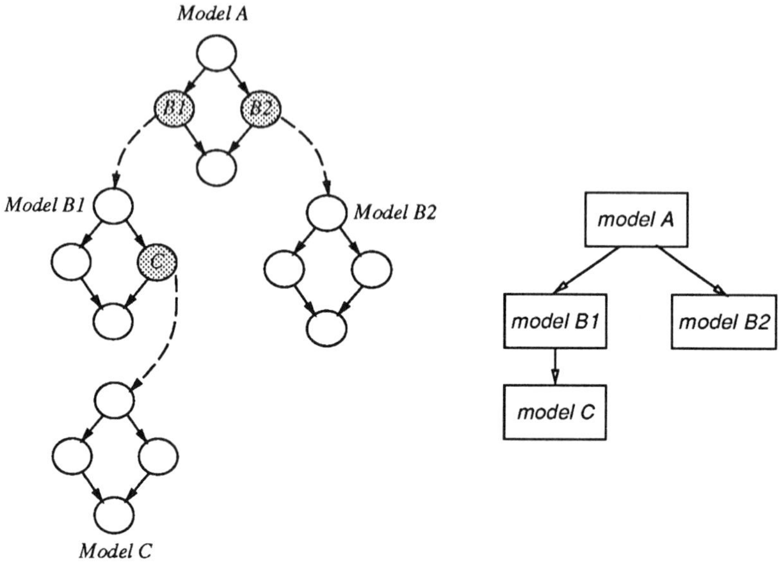

Calling hierarchy for A

Figure 4.1: Example of the calling hierarchy for model A. Model A calls models B_1 and B_2, and model B_1 calls model C.

Each vertex represents an operation requiring an integral number of control states (clock cycles), possibly zero, to execute. The *execution delay* of a vertex v_i represents the number of cycles it takes to execute. Execution delays are defined by the mapping $\delta : V \to Z^+$ from the set of vertices to non-negative integers, where $\delta(v_i) \geq 0$ denotes the execution delay of vertex v_i. These delays are derived either from the operation type, i.e. loading a register takes one clock cycle, or from estimates obtained through logic synthesis, i.e. delay is obtained by computing the critical delay through the logic expressions normalized to the cycle time.

4.1.1 Hierarchy in the model

In this section, we further separate the sequencing graph hierarchy into two forms: *calling hierarchy* and *control-flow hierarchy*.

- *Calling hierarchy: Calling hierarchy* refers to the nesting structure of procedure and function calls in the model. Consider for example four

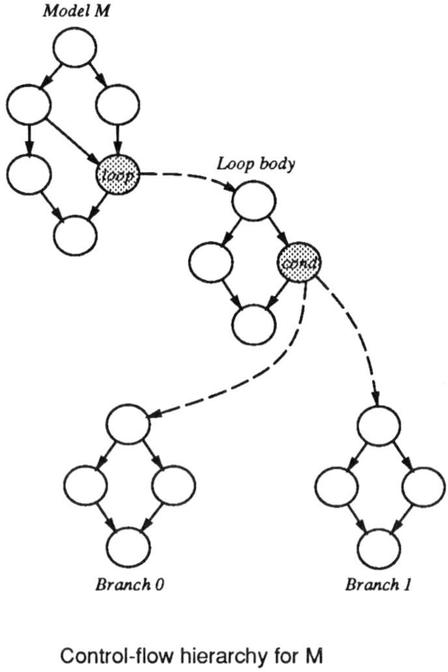

Figure 4.2: Example of the control-flow hierarchy for model M containing a loop vertex, which in turn contains a conditional vertex with two branches.

models A, B_1, B_2, and C. Suppose A calls B_1 and B_2, and B_1 in turn calls C, then the calling hierarchy for A is as shown in Figure 4.1.

- *Control-flow hierarchy: Control-flow hierarchy* refers to the nesting structure of conditionals and loops in the sequencing graph. For example, the body of a loop vertex is a sequencing graph of lower hierarchy and each branch of a conditional is a separate sequencing graph of lower hierarchy. An example of control-flow hierarchy is shown in Figure 4.2.

Let M be a process, procedure or function model, which is represented in general by a hierarchy of sequencing graphs. The sequencing graph at the root of the hierarchy is called the *root graph* of M, denoted by G_M. The *cf-hierarchy* of G_M, denoted by G_M^*, is the control-flow hierarchy of G_M. In Figure 4.2, the cf-hierarchy of model M consists of all four graphs in the figure.

The motivation for defining the two forms of hierarchy above stems from our model of hardware resources. A model in the behavioral description corresponds to a hardware resource that can be shared. Therefore, the calling hierarchy for a model defines the set of hardware resources that it depends on for synthesis; the control-flow hierarchy represents its behavior in terms of data-flow and control-flow operations. Hardware resources are described in detail in Section 4.3. In addition, the use of hierarchy to represent loops allows the sequencing graph model to be acyclic; this greatly simplifies the subsequent synthesis formulations and algorithms.

4.1.2 Data-dependent delay operations

A problem arises for conditionals or loops because their execution delays depend on external signals and events that are not known statically. We further categorize the vertices based on this observation.

Definition 4.1.2 *A vertex has* **data-independent (fixed) delay** *if the time required to execute its operation is fixed for all input data sequences. Otherwise, the vertex has* **data-dependent delay**.

The delay associated with a fixed delay vertex depends solely on the nature of the operation. Examples include addition and register loading. In contrast, the time to execute a data-dependent delay operation is represented by a range of possible delays. In some cases, this range is finite; for example, the range for a conditional vertex with one branch taking 2 cycles and the other branch taking 3 cycles is $[2, 3]$. In general, however, the range of possible delay values may be unbounded, such as for loops whose exit condition depends on some signal value, or for message passing commands that synchronize between two concurrent processes. Data-dependent delay vertices are important in specifying interfaces and handshaking protocols. A data-dependent delay vertex is called an **anchor** of the graph. Anchors are used in the relative scheduling formulation, described in Chapter 6.

This categorization is hierarchical. A sequencing graph where all vertices have fixed delay is called a *fixed delay sequencing graph*. A call to a fixed delay sequencing graph, or a conditional where all branches have the same fixed delay, or a fixed iteration loop whose body has fixed delay are all classified as having fixed delay.

A vertex is further classified according to the value of its execution delay for a particular input sequence. If a vertex requires one or more cycles to execute (execution delay > 0), then it is called a **state** vertex. Otherwise, it is called a **stateless** vertex (execution delay $= 0$). A graph with only stateless vertices is a stateless graph.

```
if ( opcode == MULTIPLY ) {
    /* multiplication requires more than 1 cycle */
    result = mult(a, b);
} else
if ( opcode == LOGIC_AND ) {
    /* simple Boolean operation */
    result = a & b;
}
```

Figure 4.3: Example of a conditional where the branches have different execution delay.

Category	*Type*	*Fixed/Data-dependent*	*Stateless/State*
simple	*no-op*	Fixed	Stateless
	computation	Fixed	Stateless or State
	logic block	Fixed	Stateless or State
	load register	Fixed	State
	input/output	Fixed	State
	send/receive	Data-dependent	State
complex	*call*	*either*	Stateless or State
	conditional	*either*	Stateless or State
	loop	*either*	Stateless or State

Table 4.2: Properties of vertices in the sequencing graph model.

For fixed delay vertices, the property of *stateless* versus *state* is fixed; e.g. a *no-op* vertex is always stateless and a *load-register* (takes one cycle) is always a state vertex. For data-dependent delay vertices, however, this property depends on the value of execution delay for a particular input sequence. For example, an un-entered while loop (exit condition is true when the loop is first executed) does not require any clock cycles and therefore is stateless for that input sequence.

Consider for example a conditional vertex with two branches, as shown in Figure 4.3. The first branch is a state sequencing graph because it requires at least one cycle to execute. The second branch is a stateless sequencing graph because it contains only combinational logic. If for a particular input sequence the selected branch is stateless, i.e. `(opcode == LOGIC_AND)`, then the conditional vertex is stateless for that input sequence. Likewise, if a new input sequence is applied to the conditional causing the other branch to be selected, i.e. `(opcode == MULTIPLY)`, then it becomes a state vertex. The same analysis

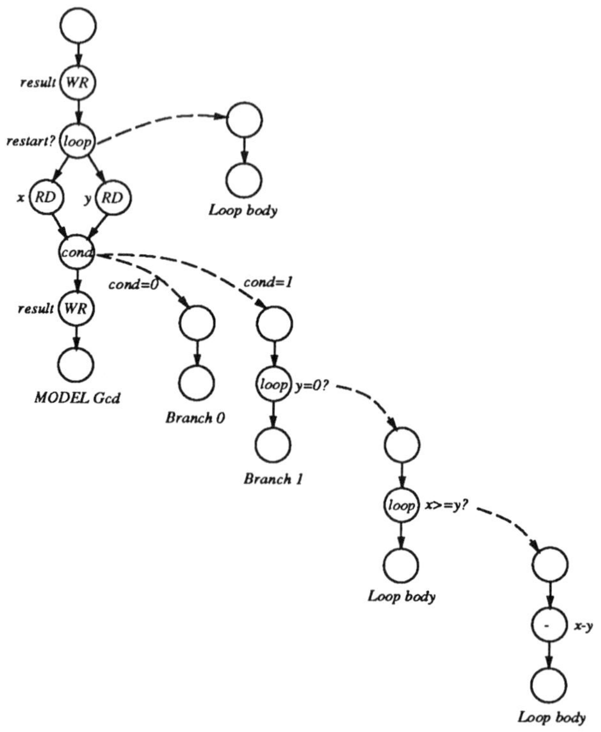

Figure 4.4: The sequencing graph for the Greatest common divisor example.

applies to call and loop vertices. Table 4.2 summarizes these properties.

4.1.3 Sequencing graph examples

The sequencing graph for the *gcd* example of Figure 2.2 is shown in Figure 4.4. Execution flow starts from the source vertex and terminates at the sink vertex. The dashed arcs represent transfer of control from a complex vertex, such as a loop or a conditional, to a sequencing graph lower in the control-flow hierarchy. The root sequencing graph *gcd* first sets the output `result` to zero, then enters a loop that exits on the rising edge of `restart`. No operation is performed in the loop body and therefore it is represented by only the source and sink vertices. Upon detecting the rising edge, the loop exits and the operands `x` and `y` are read in parallel. Euclid's iterative algorithm is modeled by a conditional node and its nested subgraphs. In particular, the `case=1` branch consists of two nested loops. Finally, the computed greatest common divisor value is written to `result`.

Figure 4.5 illustrates the sequencing graphs for the *encoder* and *decoder* processes of the error-correction example in Section 2.1.5. Shaded vertices labeled with "C" denote calls to the *PARITY* function and vertices labeled with "E" denote logic expression blocks. In the encoder process, after waiting for the rising edge of the ready signal, the data is read. It then computes the parity bits by calling *PARITY*. Once the parity bits have been computed, the output ready pulse is written in parallel with the serial transmission of the 16-bit data frame. The steps are reversed in the *decoder* process by first reading the serial data stream, then computing the parity bits to detect transmission errors. Once error correction has taken place, the output data, error flag, and the output ready pulse are written out in parallel.

4.2 Timing Constraints

The sequencing edges represent the precedence relationships that are due to data-flow and control-flow dependencies. This section describes the derivation of a *constraint graph* model from the sequencing graph model with timing constraints. The constraint graph captures the timing behavior and timing requirements of a given sequencing graph.

Consider a sequencing graph $G_s(V, E_s, \delta)$. Let $T(v_i)$ represent the *start time* of v_i, i.e. the time at which v_i begins execution with respect to the source vertex of G_s. Detailed timing constraints consist of the following:

- *Minimum* timing constraints $l_{ij} \geq 0$ from v_i to v_j, requiring that $T(v_j) \geq T(v_i) + l_{ij}$. This constraint implies that v_j should be activated at least l_{ij} cycles after the activation of v_i.

- *Maximum* timing constraints $u_{ij} \geq 0$ from v_i to v_j, requiring that $T(v_j) \leq T(v_i) + u_{ij}$. This constraint implies that v_j should be activated no more than u_{ij} cycles after the activation of v_i.

The timing behavior of a sequencing graph $G_s(V, E_s, \delta)$ under timing constraints is captured by a polar, edge-weighted, directed **constraint graph** $G(V, E, \omega)$. Vertices of the constraint graph are identical to the vertices of the sequencing graph; they represent the *activation* of the corresponding operations. Edges capture the minimum and maximum timing relationships between the activation of operations. They are categorized into *forward* (E_f) and *backward*

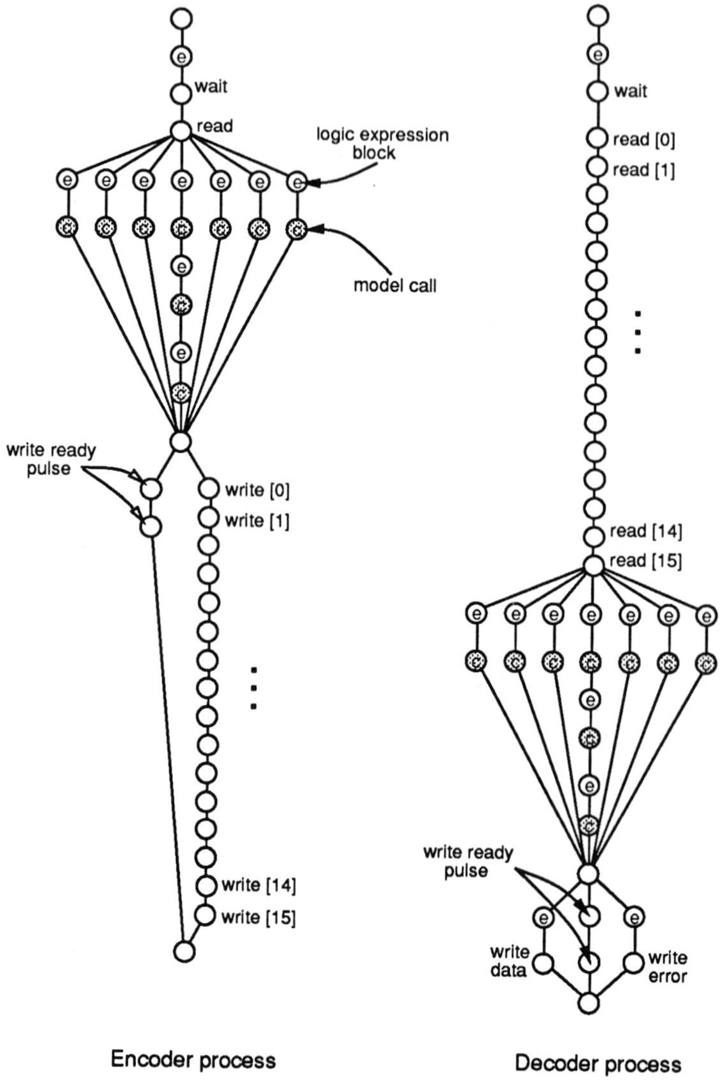

Figure 4.5: The sequencing graph for the *encoder* and *decoder* processes in the Error correcting code example.

Item	*Constraint graph*		
	Type	Edge	Edge Weight
minimum sequencing edge s_{ij}	forward	(v_i, v_j)	$\delta(v_i)$
maximum sequencing edge x_{ij}	backward	(v_j, v_i)	$-(\delta(v_i) + x_{ij})$
minimum timing constraint l_{ij}	forward	(v_i, v_j)	l_{ij}
maximum timing constraint u_{ij}	backward	(v_j, v_i)	$-u_{ij}$

Table 4.3: Edge types in the constraint graph model.

(E_b) edges, i.e. $E = E_f \cup E_b$. Weights are associated with the edges by the mapping $\omega : E \rightarrow Z$, which assigns to each edge e_{ij} a weight $\omega(e_{ij})$ that corresponds to the following inequality constraint between v_i and v_j:

$$T(v_i) + \omega_{ij} \leq T(v_j)$$

Forward edges have positive weights and represent minimum sequencing and timing constraints; backward edges have negative weights and represent maximum sequencing and timing constraints. The derivation of edges and weights from the sequencing graph and timing constraints is described below.

- *Minimum sequencing edge* $s_{ij} \in E_s$: A forward edge $e_{ij} \in E_f$ with weight $\omega(e_{ij}) = \delta(v_i)$, modeling a minimum timing constraint equal to the execution delay of v_i, e.g., v_j should start at least after the completion of v_i.

- *Maximum sequencing edge* x_{ij}: A backward edge $e_{ji} \in E_b$ with weight $\omega(e_{ji}) = -(\delta(v_i) + x_{ij})$, modeling a maximum timing constraint on v_j after the completion of v_i, e.g., v_j should start no more than x_{ij} cycles after the completion of v_i.

- *Minimum timing constraint* l_{ij}: A forward edge $e_{ij} \in E_f$ with weight $\omega(e_{ij}) = l_{ij}$.

- *Maximum timing constraint* u_{ij}: A backward edge $e_{ji} \in E_b$ with weight $\omega(e_{ji}) = -u_{ij}$, because $T(v_j) \leq T(v_i) + u_{ij}$ can be rewritten as $T(v_j) - u_{ij} \leq T(v_i)$.

Figure 4.6 and Table 4.3 summarize the types of timing constraints. Analysis techniques for consistency of these timing constraints is the topic of Chapter 6.

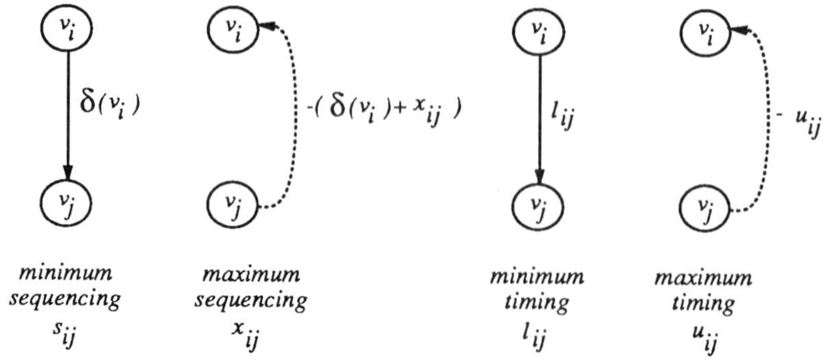

Figure 4.6: Types of sequencing and timing constraints.

Since the execution delays can be data-dependent, it is clear that the edge weights can also be data-dependent due to the presence of sequencing constraints. In particular, if operation v_i has data-dependent delay $\delta(v_i)$, then a minimum sequencing constraint s_{ij} will yield a forward edge with data-dependent weight $\delta(v_i)$. Likewise, a maximum sequencing constraint x_{ij} will yield a backward edge with data-dependent weight $-(\delta(v_i) + x_{ij})$. The need for data-dependent backward edges arises when we would like to constrain an operation v_j to begin execution at most a certain number of cycles *after* the completion of another operation v_i. If v_i has data-dependent delay, then such constraint cannot be modelled by a fixed maximum timing constraint. We assume that any maximum sequencing constraint (and hence its derived data-dependent backward edge) x_{ij} is matched with a corresponding minimum sequencing constraint s_{ij}.

Constraint graph examples. An example of a constraint graph is given in Figure 4.7, where s is the source, t is the sink, and double circles represent operations with data-dependent delay. The operation c must begin at least 3 cycles and at most 5 cycles after operation b begins execution, which is specified by the equation $T(b) + 3 \leq T(c) \leq T(b) + 5$. Similarly, operation d must begin at least 3 cycles and at most 5 cycles after the *completion* of operation b, given by the equation $T(b) + \delta(b) + 3 \leq T(d) \leq T(b) + \delta(b) + 5$. In this case $\delta(b)$ represents the execution delay of operation b, where the delay is not fixed. For example, b could represent a *wait* operation on an external synchronization signal.

Finally, Figure 4.8 shows the constraint graph derived from the sequencing graph of the *encoder* process in Figure 4.5. The execution delay of an expression

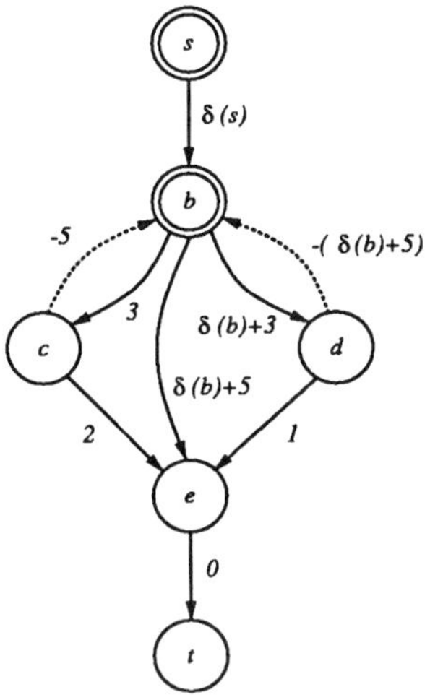

Figure 4.7: Example of a constraint graph with minimum and maximum constraints.

block e is denoted by $d(e)$. The execution delay of a call to *PARITY* is denoted by $d(c)$. Reading and writing of ports require 1 cycle to execute. Although there are no timing constraints other than those derived from the sequencing dependencies, they can be derived in a straightforward manner using the derivation described above.

4.2.1 Properties of constraint graph

Without loss of generality, the subgraph $G_f = (V, E_f)$ induced by forward edges is assumed to be *acyclic*, i.e. a minimum timing constraint l_{ij} is not considered to be valid if there is already a path of dependencies from v_j to v_i. In particular, if $l_{ij} > 0$, then the constraint violates the dependencies between the operations; otherwise, if $l_{ij} = 0$, then the constraint can be modeled equivalently by a maximum timing constraint $u_{ji} = 0$ from v_j to v_i. Cycles in the forward constraint graph can be detected with Dijkstra's algorithm [BN86]. The values of the execution delays are irrelevant for this check.

A vertex v_i is a predecessor of vertex v_j ($v_i \in pred(v_j)$) if there is a directed path in $G_f = (V, E_f)$ from v_i to v_j. Conversely, a vertex v_i is a successor of vertex v_j ($v_i \in succ(v_j)$) if there is a directed path from v_j to v_i. The length of the longest weighted path from v to w in the full graph $G(V, E)$ is denoted

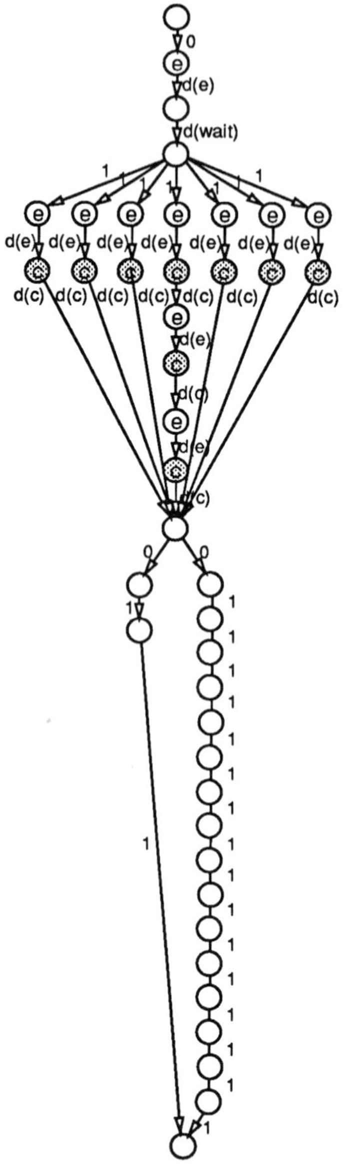

Constraint graph for Encoder

Figure 4.8: The constraint graph model for the *encoder* process of the Error correcting code example.

by $length(v, w)$, where all data-dependent edge weights are set to zero.

The execution delay of each vertex must be associated with *a particular input sequence* when there are vertices in the graph with data-dependent delay. The reason is because the time required to achieve synchronization and the number of iterations necessary for a data-dependent loop are known only in the context of a given input sequence.

The **latency** of a constraint graph is the minimum number of cycles that is required to execute all operations subject to timing constraints. The latency is computed hierarchically in a bottom-up manner according to the following definition.

Definition 4.2.1 *For a particular input sequence, the* **latency** *of a constraint graph is equal to the length of the longest weighted path from source to the sink, where:*

1. *The execution delay of a call vertex is exactly equal to the latency of the called graph.*

2. *The execution delay of a conditional vertex is equal to the latency of the selected branch.*

3. *The execution delay of a loop vertex is equal to the number of iterations multipled by the latency of the loop body.*

From the above definition, $length(v_0, v_N)$ is a lower bound for the latency:

$$length(v_0, v_N) \leq latency(G)$$

4.3 Hardware Resources

The data-path in the final hardware implementation consists of three types of elements: functional units, registers, and multiplexers. Functional units correspond to arithmetic operations (e.g. $+$ or $*$) or to generic models (e.g. a procedure describing some application-specific functions). Registers are introduced either by the input behavioral specification or as required to implement hardware sharing. Multiplexers form the interconnect logic to steer appropriate signals between functional units and registers.

Whereas the designer has control over the allocation and binding of operations to functional units, register and multiplexer elements are created automatically by the system. Once the designer has specified a particular resource binding configuration, registers and multiplexers are introduced to implement this binding. Therefore, as far as the designer is concerned, the only controllable hardware resources are functional units corresponding to the called procedures and functions.

Section 4.3.1 describes the correspondence between hardware functional u-nits and models in the behavioral description. Registers and multiplexer are described in Sections 4.3.2 and 4.3.3, respectively.

4.3.1 Functional units

This research treats each *model* in the input description as a *resource* that can be shared among the calls to that model. Each different implementation of the called model represents a specific *resource type*, which has its own area and performance characteristics. Predefined operators, such as $+$ or $-$, can be converted into calls to the appropriate library models (using user-driven behavioral transformations), but are by default realized as combinational logic. Due to the size of combinational multiplication and division, all multiply ($*$) and divide ($/$) operators are always mapped into calls to library models. Therefore, the only operations whose implementing hardware can be allocated and controlled by the designer are calls to procedure or function models. This model of resources implies that resource sharing is possible only for *call* vertices in the sequencing graph model.

There are several motivations for treating models and resources in this manner. First, since many complex ASIC designs use application-specific logic functions to describe hardware behavior, the delay and area attributes of these modules are not known *a priori* since they depend on the particular details of the logic functionality. Having the ability to synthesize each model in a bottom-up fashion according to its distinct needs allows the calling models to more accurately estimate their resource requirements. Second, the granularity of resource sharing can be controlled by the designer by modifying the calling hierarchy in the high level specification. Finally, instead of relying on parametrized and predefined modules, logic synthesis techniques applied hierarchically to each model can significantly improve the quality of the resulting design.

For example, two calls to a model A can be implemented either by a single resource corresponding to a hardware implementation of A, where both calls share the use of the resource, or by two resources, where each call is implemented by a different resource. The first case is shown in Figure 4.9(a) and the second case is shown in Figure 4.9(b). Note that when resources are shared, it is

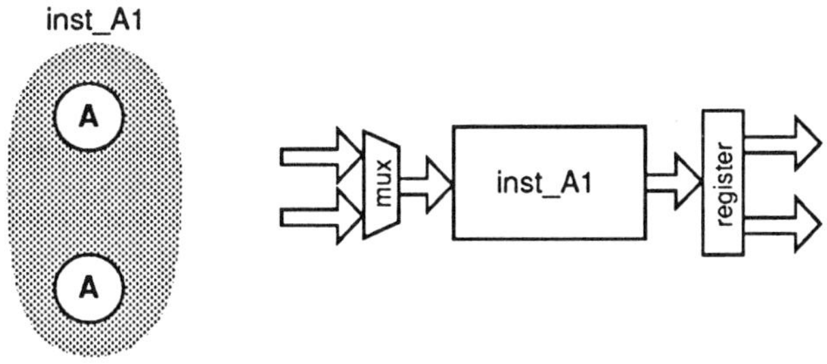

(a) Both calls share one resource inst_A1

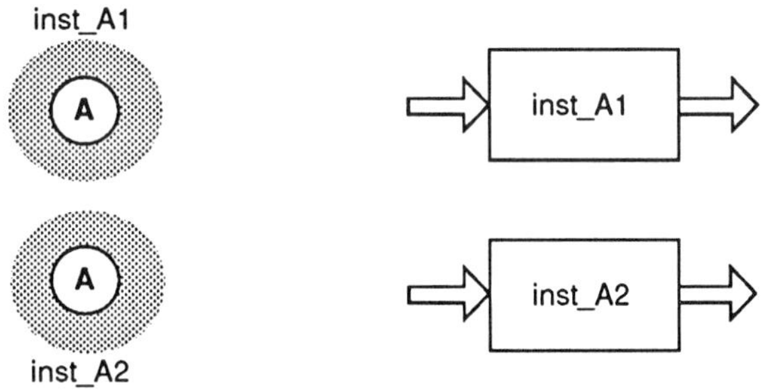

(b) Each call is dedicated to a resource instance

Figure 4.9: Illustrating the alternatives in implementing two calls to model A: (a) both calls share one resource, (b) each call is implemented by separate resources.

necessary to introduce multiplexing logic to steer the appropriate values to the input ports of the resource component and registers to latch the output of the resource component. Control logic must ensure that the two calls to A do not execute simultaneously. The allocation of registers and multiplexers to support hardware sharing is described in the following sections.

4.3.2 Registers

The registers in the data-path are categorized into *user-defined* and *temporary* registers. User-defined registers are derived from the input description by one of the following methods. First, registers can be explicitly declared as static variables in HardwareC. Second, variables that are referenced before assigned in data-dependent loops are identified by the reference stack algorithm in Hercules and implemented as registers. Finally, registers are used to implement the input and output ports, i.e. a register is associated with each output port, and each write to the port is implemented as updating the register with the appropriate value.

Temporary registers hold internal values that do not correspond to user-defined variables. They are created as a result of hardware sharing. If a resource component is shared among multiple operations, then the outputs of the resource component may have to be *latched* so that the results of each different invocation of the resource component are retained even if the component is reused subsequently.

Consider a sequence of k operations in a sequencing graph, $v_1 \rightarrow v_2 \rightarrow \cdots \rightarrow v_k$, all bound to the same resource component. In general, it is necessary to latch the outputs of the component for every operation in the sequence, since the intermediate results for $\{v_i | i = 1, \cdots, k\}$ may be referenced after the resource component has been released and hence must be latched for future accesses. Note that the latch is separated from the resource because it depends on the context in which the resource is called. Figure 4.10 illustrates the concept.

The execution delay for an operation whose hardware resource is latched must be increased by one to account for the clock cycle required to latch the outputs. The delay $\delta(v)$ for a vertex v using a resource component r with delay $delay_r$ is:

$$\delta(v) = \begin{cases} delay_r & \text{no latching} \\ delay_r + 1 & \text{latching} \end{cases}$$

For example, if a combinational logic function is shared, then the latched function call requires 1 cycle. The call delay is computed hierarchically, e.g., if the latched call to a combinational logic function represents another shared function, then each latched call will require 2 cycles. Since the cascaded latching is not necessary in general, optimizations can be performed to remove the redundant latches. Such optimizations were not implemented in the current version of Hebe.

Latching requirements are determined according to the semantics of the sequencing graph model. No assumptions are made on whether the latching register is folded or dedicated. Register folding (i.e. sharing of registers) to minimize the number of registers in the design can be applied after resource binding and conflict resolution have been performed. Variable lifetime is used to determine when two registers can be folded. The folding algorithm is based on a variant of the left-edge channel routing algorithm similar to REAL [KP87].

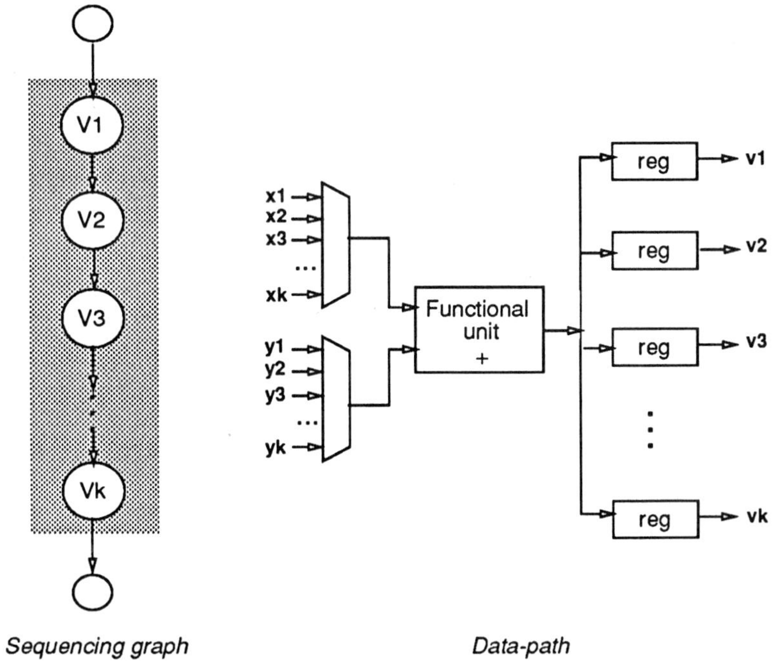

Figure 4.10: Example of latching outputs for (a) a sequencing graph containing a sequence of addition operations, all bound to the same resource component, and (b) the corresponding latching requirements.

4.3.3 Multiplexers

The *interconnect* structure steers signals between hardware components in the data-path. If there is no hardware sharing and all operations are implemented by dedicated functional units, then the interconnect reduces to direct wiring between ports of the functional units. Otherwise, a selection structure is required and control logic is defined to ensure contention-free execution. For example, if two addition operations use the same adder, then the inputs to the adder are multiplexed values of the operands of these two operations. The selection of which operand to use depends on the schedule.

The synthesis flow in Hebe performs resource binding before scheduling. Each resource binding can be evaluated because the interconnect structure is *completely* specified for a particular binding configuration. Once operations are completely bound to resources, the subsequent scheduling phase will affect only the control signals that are used by the interconnect, without modifying its structure. Therefore, evaluations on the area and delay of the interconnect can be performed *before* scheduling.

Hebe implements the interconnect structure by a *multiplexer-based* scheme. However, instead of using predefined library components, the multiplexer structure is described as logic expressions that are subsequently optimized by logic synthesis.

4.4 Summary of Sequencing Graph Model

The sequencing graph model is a simple yet powerful representation for synthesis. Both control-flow and data-flow dependencies are captured uniformly as edges in the sequencing graph. The main features of this model are summarized below.

- *Hierarchy.* This model uses hierarchy in two ways: a calling hierarchy that represents the nesting of procedure and function calls, and a control-flow hierarchy that represents conditional branching and loops. As a result, each graph in the model is acyclic.

- *Data-dependent delay operations.* The execution delay of a vertex can be fixed or data-dependent. The latter type describes external synchronizations and loops.

- *Concurrency.* Each directed path through the sequencing graph corresponds to a thread of execution flow. This model supports multiple threads of concurrent execution flow.

- *Detailed timing constraints.* The constraint graph model that is derived from a sequencing graph under timing constraints is used to capture the timing behavior and timing requirements of a design.

The sequencing graph model is the underlying representation for design space exploration, which is described in the next chapter. Relative scheduling, constrained conflict resolution, and relative control synthesis and optimization are all formulated based on the constraint graph model.

Chapter 5

Design Space Exploration

If it was practical to use as many hardware resources as operations to synthesize a given behavior, then each operation could be implemented by a dedicated hardware resource. In this case, hardware sharing dependencies would not exist and the operation execution order would only be constrained by data dependencies and synchronizations. However, the assumption of dedicated hardware resources is often not appropriate for large-scale designs for two reasons. First, the size of the dedicated implementation would typically make it impractical. Second, the interconnection delays would probably not be negligible, and therefore the resulting hardware would not be the fastest achievable implementation in practice.

Hebe performs *structural synthesis* to explore the design implementation tradeoffs through hardware resource sharing, while ensuring the resulting implementation satisfies the user specified constraints on resource and timing. The spectrum of possible implementation alternatives is called the *design space* for the behavior. Section 5.1 describes the formulation of the design space as the set of possible resource bindings. The concept of *concurrency factor*, which is used extensively in both resource allocation and design space exploration, is introduced in Section 5.2. Section 5.3 describes search strategies to explore the design space. Hebe supports both exact and heuristic search strategies. Using the exact strategy, the designer can find an optimal implementation with respect to area and or performance. For designs with a large design space, a heuristic exploration strategy is used to synthesize a portion of the total design space. Section 5.4 summarizes the design space formulation and exploration strategy in Hebe.

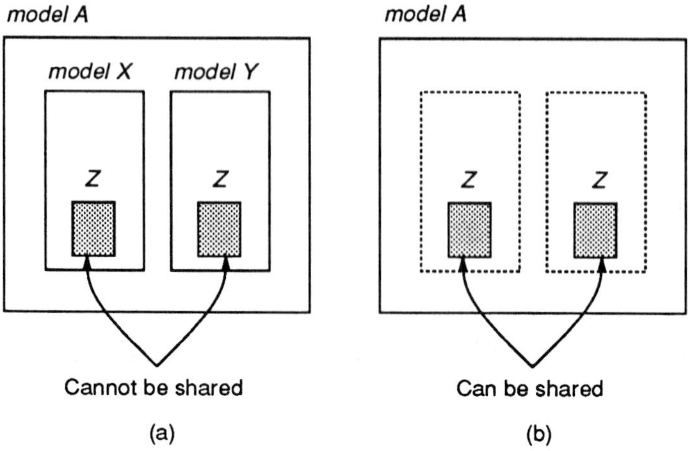

Figure 5.1: Resource sharing across the call hierarchy: (a) model Z cannot be shared among models X and Y, (b) after flattening, model Z can be shared among the two invocations in model A.

Synthesis across the hierarchy. As described in Chapter 4, the sequencing graph model supports two forms of hierarchy: *calling hierarchy*, which refers to the nesting structure of model call vertices, and *control-flow hierarchy*, which refers to the nesting structure of conditionals and loops.

All models that are called in the control-flow hierarchy of a model must already be synthesized before the given model can be considered for synthesis. Based on this bottom-up synthesis strategy, resource sharing across the calling hierarchy is not allowed. To illustrate this point, consider the example in Figure 5.1. Model A calls models X and Y, each of which, in turn, call model Z. The resource implementing Z cannot be shared between the two calls in X and Y. However, the designer can control the granularity of resource sharing by selectively performing *in-line expansion* of model calls during behavioral synthesis in Hercules (see Chapter 3). Returning to the above example, the designer can flatten the calls to X and Y in model A in order to expose the calls to Z on the same level; once flattened, sharing can take place. Whereas resource sharing is not supported across the calling hierarchy, it *is* supported across the control-flow hierarchy. Resource sharing across the control-flow hierarchy is the topic of this chapter.

Recall from Section 4.3 that a call vertex in a sequencing graph corresponds to an invocation of a resource component. Each call can be implemented by one or more resource types, where each type represents a particular hardware implementation of the called model. An important assumption in Hebe is that the resource type implementing each call vertex is specified prior to structural synthesis. Hebe will perform tradeoffs in the *number of* resources that are allocated, not in the *types of* resources implementing the operations.

5.1 Design Space Formulation

Given a sequencing graph, different hardware implementations may exist. Each implementation corresponds to a particular configuration of hardware resources and the appropriate control logic. This section presents the formulation of the *design space*. In Section 5.1.1, we present as background the concepts of resource allocation and binding based on the graph model. The formulation of the design space is then described in Section 5.1.2.

5.1.1 Background

For a sequencing graph $G_s(V, E_s, \delta)$, we focus our attention on the subset of vertices in G_s^* (cf-hierarchy of G_s) whose corresponding resources can be shared. These operations are called *shareable operations* and are denoted by $\widehat{V} \subseteq V^*$, where V^* represents all vertices belonging to graphs in G_s^*. Recall from Section 4.3 that resources implementing *call* vertices can be shared and controlled by the designer. Therefore, $\widehat{V}$ is defined to be the set of *call* vertices in G_s^*. Non-shareable operations are not considered because their implementing hardware is dedicated and not controllable by the designer. Shareable operations define the scope within which resource allocations and bindings are defined.

Consider, for example, the sequencing graph hierarchy of Figure 5.2. The root graph G_s contains 2 calls to model A and 1 call to model B. It also contains a loop, the body of which is a sequencing graph containing two call vertices: one to A and the other to B. The set of shareable operations is $\widehat{V} = \{A1, A2, A3, B1, B2\}$.

Let $\mathcal{T}$ denote the set of resource types in $\widehat{V}$. For example,

$$\mathcal{T} = \{ \text{ model A, model B } \}$$

the set $\mathcal{T}$ represents the resource types for the example in Figure 5.2. The *operation set* for type $t \in \mathcal{T}$, denoted as $O(t) \subseteq \widehat{V}$, consists of shareable

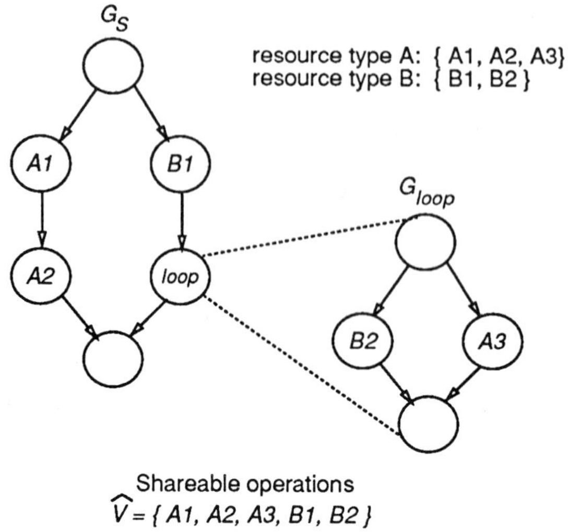

Figure 5.2: Example of sequencing graph with 3 calls to model A and 2 calls to model B, where $\widehat{V} = \{A1, A2, A3, B1, B2\}$.

operations with resource type t. A *resource allocation* is formally defined as follows.

Definition 5.1.1 *Given a sequencing graph $G_s(V, E_s, \delta)$ and resource types $\mathcal{T}$, a* **resource allocation**[1] *is the mapping $\alpha : \mathcal{T} \rightarrow Z^+$ from the set of resource types to positive integers, where $\alpha(t)$ denotes the number of resources allocated for resource type $t \in \mathcal{T}$.*

Each resource type in $\mathcal{T}$ must have at least one resource, i.e. $\alpha(t) \geq 1\ \forall t \in \mathcal{T}$, since otherwise an implementation is not possible. A resource instance in an allocation α is described by a pair (t, i), where $t \in \mathcal{T}$ denotes the type of the resource instance and i $(1 \leq i \leq \alpha(t))$ denotes the specific allocated instance. For example, Figure 5.3 shows a resource allocation for the sequencing graph example in Figure 5.2: 2 instances of model A ($\alpha(\text{A}) = 2$) and 1 instance of model B ($\alpha(\text{B}) = 1$). The range of possible allocations for model A is $1 \leq \alpha(A) \leq 3$ and for model B it is $1 \leq \alpha(B) \leq 2$.

Given a resource allocation α, a *resource binding* for a sequencing graph G_s is an assignment of shareable operations $\widehat{V}$ to specific instances of the allocated

[1]The term allocation has often been overloaded in the literature, e.g., some authors refer to allocation as binding. We will strictly adhere to our definition in the sequel.

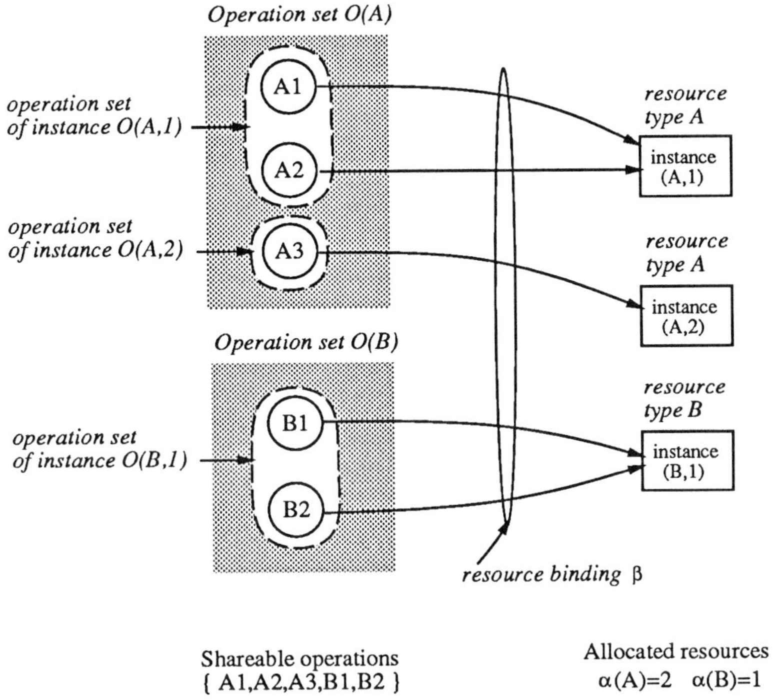

Figure 5.3: Illustrating the relationship between shareable operations and allocated resources. The allocation is $\alpha(A) = 2$ and $\alpha(B) = 1$, and the arcs represent the resource binding β.

resources. It is defined as follows.

Definition 5.1.2 *A* **resource binding** *of a sequencing graph G_s given a resource allocation α is a mapping $\beta : \widehat{V} \rightarrow (\mathcal{T} \times Z^+)$, where $\beta(v) = (t, i)$ if operation $v \in \widehat{V}$ is being implemented by the i-th instance of resource type $t \in \mathcal{T}$, $1 \leq i \leq \alpha(t)$; otherwise, $\beta(v)$ is undefined.*

A vertex for which β is defined is called a *hardware-bound* vertex; otherwise it is called an *hardware-unbound* vertex. If there are no hardware-unbound vertices in $\widehat{V}$, then β is a *complete binding*; otherwise, it is a *partial binding*. Figure 5.3 shows a binding β defined on the sequencing graph example of Figure 5.2 for the allocation $\{\alpha(A) = 2, \alpha(B) = 1\}$.

Examples of different resource bindings for a sequencing graph containing 4 calls are shown in Figure 5.4 (b) through (e). All operations grouped by the

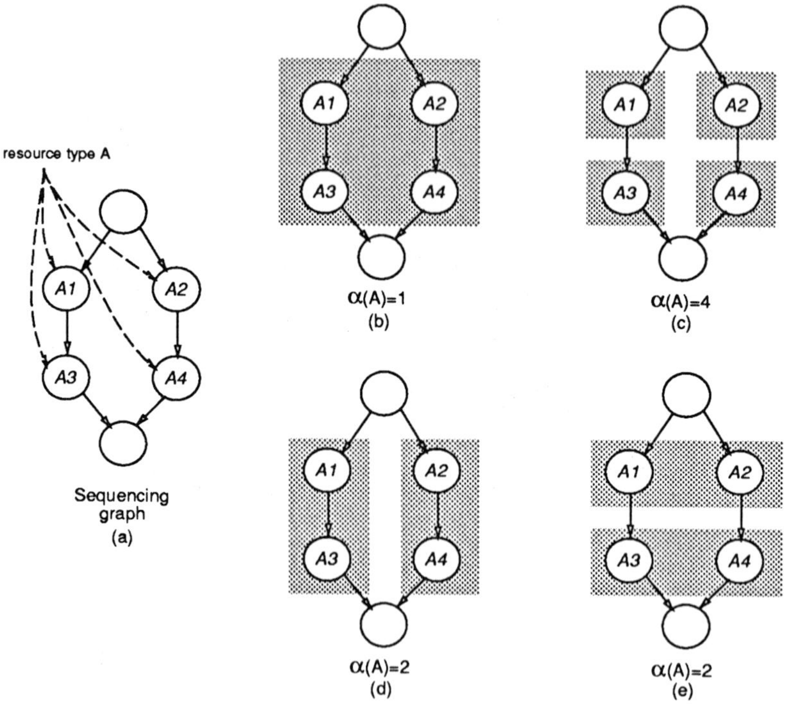

Figure 5.4: Examples of different resource bindings, where operations within a shaded block are bound to the same resource instance.

shaded rectangle share the same hardware resource in the final implementation, e.g. the binding of (b) utilizes one resource, the binding of (c) utilizes four resources.

A partial binding can be defined for more than one allocation, where it is assumed that the number of resources required by the partial binding is satisfied by these allocations. This leads to the concept of *compatible* bindings, defined below.

Definition 5.1.3 *A complete binding β_c is* **compatible** *with a partial binding β_p for a resource allocation α, denoted by $\beta_c \overset{\alpha}{\prec} \beta_p$, if for all hardware-bound vertices $v \in \widehat{V}$, the implementing resource instance is identical: $\beta_p(v) = \beta_c(v)$.*

In other words, a compatible binding can be derived from a given partial binding by mapping all hardware-unbound vertices to resource instances. Obviously,

Symbol	*Name*	*Description*
G_s	*root graph*	Root of sequencing graph hierarchy
G_s^*	*cf-hierarchy of G_s*	Control-flow hierarchy of G_s
V^*	*operation domain of G_s*	All vertices in cf-hierarchy G_s^*
$\widehat{V}$	*shareable operations of G_s*	All call vertices in V^*
$\mathcal{T}$	*resource type set*	Set of all resource types for G_s
$O(t)$	*operation set of t*	Subset of $\widehat{V}$ with type $t \in \mathcal{T}$
$\alpha(t)$	*resource allocation for t*	# of allocated instances of type $t \in \mathcal{T}$
(t,i)	*resource instance*	i^{th} instance of type $t \in \mathcal{T}$
β_P	*partial resource binding*	Partial mapping of $\widehat{V}$ to α
β_C	*complete resource binding*	Complete mapping of $\widehat{V}$ to α
$O_{(t,i)}$	*instance operation set of (t,i)*	All vertices bound to instance (t,i)

Table 5.1: Summary of resource allocation and binding terminology.

if β_p is already a complete binding (i.e. all operations are pre-assigned to resources), then there is a single compatible binding. From now on, we describe a (compatible) binding by β without the subscript, when there is no confusion with a partial binding.

Each resource instance (t,i) is bound to a subset of vertices $O_{(t,i)} \subseteq \widehat{V}$ called the **instance operation set** of (t,i), i.e. $O_{(t,i)} = \{v|\beta(v) = (t,i)\}$. The cardinality of $O_{(t,i)}$ is denoted by $|O_{(t,i)}|$. Instance operation sets partition $\widehat{V}$ into groups, each of which is implemented by a particular allocated resource instance. Obviously, an instance operation set of (t,i) is a subset of the operation set of t, i.e. $O_{(t,i)} \subseteq O(t)$, and the union of the instance operation sets for all allocations of t is equal to the operation set of t, i.e. $\cup_{i=0}^{\alpha(t)} O_{(t,i)} = O(t)$.

If there is a single instance allocated for a particular resource type t, then all operations with resource type t are automatically bound to that instance. If there is a single resource type t in the graph, then t is implied and the instance operation set is abbreviated as O_i. For example, instance operation sets for a binding β are shown in Figure 5.3. In Figure 5.4(b), there is a single instance operation set $\{A_1, A_2, A_3, A_4\}$; in Figure 5.4(c), there are four operation sets $\{A_1\}, \{A_2\}, \{A_3\}, \{A_4\}$. The definitions presented in this section are summarized in Table 5.1.

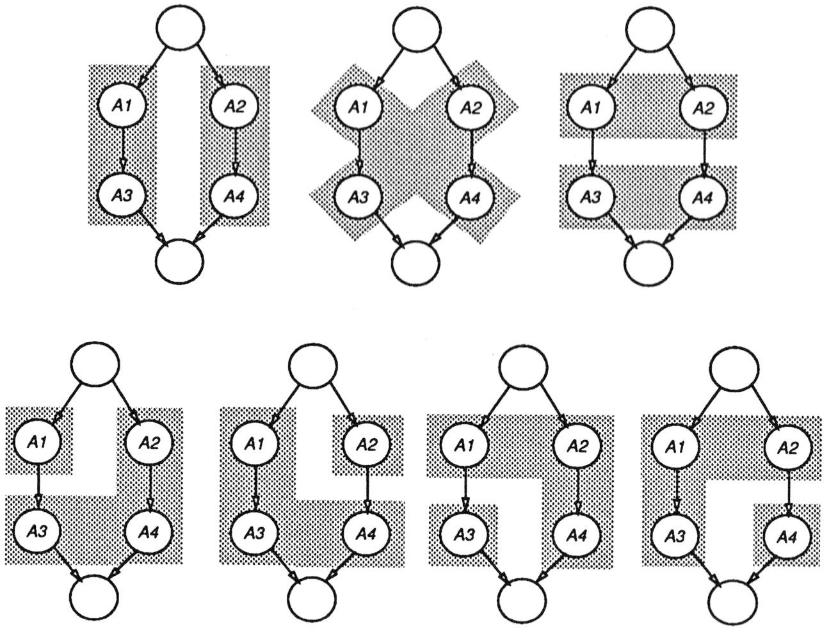

Figure 5.5: The design space for an allocation of 2 resources.

5.1.2 Defining the design space

Given resource constraints in the form of a partial binding β_p and a set of resource allocations $\{\alpha_1, \ldots, \alpha_k\}$, the *design space* of a sequencing graph $G_s(V, E_s, \delta)$ is defined as follows.

Definition 5.1.4 *For a set of resource allocations* $\{\alpha_1, \ldots, \alpha_k\}$ *and a partial binding* β_p*, the* **design space** $\mathcal{S}$ *of* G_s *is the entire set of possible compatible bindings, i.e.* $\mathcal{S} = \{\beta_c | \beta_c \stackrel{\alpha_i}{\precsim} \beta_p, \ \forall \alpha_i\}$.

The design space of possible resource bindings for Figure 5.4 with allocation $\alpha = 2$ is illustrated in Figure 5.5. There are seven different resource bindings in the design space.

An important aspect of the design space formulation is that it is a *complete* characterization of the entire set of possible design tradeoffs for a given allocation of resources. This formulation allows partial binding information to be uniformly incorporated, where the partial binding is used to limit the design space so that the synthesis system focuses on the remaining unmapped operations. At the extreme, if all operations are bound initially, then the design space trivially reduces to a single point.

Size of design space

The size of the design space influences the strategy for design space exploration. By knowing the design space size, the designer can decide whether or not exhaustive generation and synthesis of the design space is feasible. We present in this section a closed-form formula to compute the size of the design space.

Given a resource type $t \in \mathcal{T}$ and an allocation $\alpha(t)$, the problem is to compute the number of possible bindings of $|O(t)|$ operations to $\alpha(t)$ resources. $O(t)$ is the operation set of t. We assume the condition $1 \leq \alpha(t) \leq |O(t)|$ holds because otherwise the allocation is invalid. Without loss of generality, we consider the design space for *one resource type only*. The extension to support multiple resource types is to form a Cartesian product among the bindings of each resource type. We omit the (t) suffix from $O(t)$ and $\alpha(t)$ in the sequel without ambiguity.

The computation is divided into two phases. The first phase identifies a set of *partitions*, each of which is a different partition of $|O|$ objects into α ordered blocks [Aig79]. For example, if we had 3 objects ($|O| = 3$) and 2 blocks ($\alpha = 2$), then there are two possible partitions: (2,1) and (1,2). Partition (2,1) means that two objects are in the first block and one object is in the second block; partition (1,2) means that one object is in the first block and two objects are in the second block. The number of possible partitions $M(|O|, \alpha)$ is defined recursively as follows:

$$M(K,N) = \begin{cases} 1 & \text{if } K = N \\ 1 & \text{if } N = 1 \\ \sum_{i=N-1}^{K-1} M(i, N-1) & \text{otherwise} \end{cases}$$

The formula $M(K, N)$ computes the number of partitions of K operations into N resources. Three cases are possible:

- $K = N$: If the number of operations K is equal to the number of resources N, then the only feasible partition corresponds to the trivial case where each operation is assigned to a separate resource.

- $N = 1$: When only one resource is available, then the only partition is to assign all operations to the sole resource.

- *Otherwise:* The number of partitions is computed recursively by summing the following. First assign 1 operation to the first resource, then compute the number of partitions of the remaining $i = K - 1$ operations to the remaining $N - 1$ resources; next assign 2 operations to the first resource, then compute the number of partitions of the remaining $i = K - 2$ operations to the remaining $N - 1$ resources, and so on, until $i = N - 1$.

Examples of possible partitions for different values of $|O|$ and α are shown in Table 5.2.

Now consider one of the possible partition $M^i, i = 1, \cdots, M(|O|, \alpha)$ of $|O|$ operations to α resources. Let the jth block of the partition M^i be denoted as M^i_j, where the number of blocks in any partition corresponds to the number of allocated resources α. Note that each block M^i_j consists of at least one operation O, i.e. $|M^i_j| \geq 1$. The total number of operations in all blocks in M^i is exactly equal to $|O|$ due to conservation:

$$\sum_{j=1}^{\alpha} |M^i_j| = |O|$$

For each partition M^i there are α groups, where a block in this context is identical to an *instance operation set* defined earlier. The O operations are assigned to these blocks in turn. The number of possible assignments is computed as: the number of ways of selecting $|M^i_1|$ objects from $|O|$ objects: $\binom{|O|}{|M^i_1|}$, times the number of ways of selecting $|M^i_2|$ objects from $|O| - |M^i_1|$ objects: $\binom{|O| - |M^i_1|}{|M^i_2|}$, times the number of ways of selecting $|M^i_3|$ objects from $|O| - |M^i_1| - |M^i_2|$ objects: $\binom{|O| - |M^i_1| - |M^i_2|}{|M^i_3|}$, and so on until the last block M^i_α.

Since the order of assignment is not important, this product is divided by $\alpha!$. Therefore, the following quantity represents the number of bindings for a given partition M^i:

$$\frac{\binom{|O|}{|M^i_1|} \cdot \binom{|O| - |M^i_1|}{|M^i_2|} \cdots \binom{|O| - \sum_{j=1}^{\alpha-1} |M^i_j|}{|M^i_\alpha|}}{\alpha!}$$

An analytical expression for the total number of possible resource bindings $P(|O|, \alpha)$ for $|O|$ operations and α allocated instances is the sum of the above expression over the entire set of partitions, as described below.

$\|O\|$	α	*Partitions*	$M(\|O\|,\alpha)$	$P(\|O\|,\alpha)$
1	1	(1)	1	1
2	1	(2)	1	1
2	2	(1,1)	1	1
3	1	(3)	1	1
3	2	(1,2) (2,1)	2	3
3	3	(1,1,1)	1	1
4	1	(4)	1	1
4	2	(1,3) (2,2) (3,1)	3	7
4	3	(1,1,2) (1,2,1) (2,1,1)	3	6
4	4	(1,1,1,1)	1	1
5	1	(5)	1	1
5	2	(1,4) (2,3) (3,2) (4,1)	4	15
5	3	(1,1,3) (1,3,1) (3,1,1)		
		(1,2,2) (2,1,2) (2,2,1)	6	25
5	4	(1,1,1,2) (1,1,2,1) (1,2,1,1) (2,1,1,1)	4	10
5	5	(1,1,1,1,1)	1	1

Table 5.2: Examples of partitions and the number of bindings for different values of $|O|$ operations and α resource allocation.

$$P(|O|,\alpha) = \sum_{i=1}^{M(|O|,\alpha)} \left[\frac{\binom{|O|}{|M_1^i|} \cdot \binom{|O|-|M_1^i|}{|M_2^i|} \cdots \binom{|O|-\sum_{j=1}^{\alpha-1}|M_j^i|}{|M_\alpha^i|}}{\alpha!} \right]$$

The value of P for different values of $|O|$ and α are given in Table 5.2. The size of the design space for a single resource type t with allocation α is $|S| = P(|O|,\alpha)$. Note that the size remains low for modest values of $|O|$ and α.

5.2 Concurrency Factor

Before describing the design space exploration strategy, we introduce first the concept of *concurrency factor* to measure of the degree of parallelism among subsets of shareable operations $Q \subseteq \widehat{V}$. This concept is used extensively in resource allocation and heuristic exploration of the design space.

Let $G_M(V, E, \delta)$ be the sequencing graph[2] for a model M. We define the *concurrency factor* for a subset of shareable operations $Q \subseteq \widehat{V}$ as follows.

Definition 5.2.1 *Given a cf-hierarchy G_M^*, the* **concurrency factor** *of a subset of shareable operations $Q \subseteq \widehat{V}$ is the maximum number of vertices in Q that can be executing simultaneously in G_M^*. It is denoted by $cfactor(G_M, Q)$.*

The concurrency factor can range in value from 1 to $|Q|$. If the factor is equal to 1, then elements of Q must be temporally disjoint. If the factor is equal to $|Q|$, then elements of Q can all execute in parallel.

In the remainder of this section, we describe the computation strategy for the concurrency factor. Section 5.2.1 considers the simple case where G_M^* consists of a single graph, and Section 5.2.2 considers the general case where G_M^* consists of a hierarchy of graphs. For both cases, we present polynomial-time algorithms to compute the concurrency factor. We conclude this section by describing the application of concurrency factor to resource allocation in Section 5.2.3.

5.2.1 Computation for flat graphs

This section describes the strategy to compute the concurrency factor assuming the cf-hierarchy G_M^* consists of a single sequencing graph $G_M(V, E, \delta)$. In other words, no conditionals or loops exist in the graph. Since there is a single graph, the elements of Q are vertices of G_M, i.e. $Q \subseteq V$. Figure 5.6(a) shows a sequencing graph with four operations $Q = \{v_1, v_2, v_3, v_4\}$. Four different execution scenarios are shown in (b), where the dotted line passes through operations that are currently executing. Since at most two operations can be active at any time, the concurrency factor of Q is equal to 2.

The concurrency factor of Q can be computed by first constructing an undirected *disjoint compatibility graph*. Vertices of the disjoint compatibility graph, denoted by $G_Q = (Q, E_Q)$, correspond to elements of Q. Undirected edges E_Q indicate when vertices are temporally disjoint, i.e. they cannot execute in parallel.

In traditional approaches, two vertices are disjoint if they are scheduled into different control steps. However, since our model supports data-dependent delay operations, these approaches cannot be used in general. Assume all operations have unbounded execution delay, then two operations are compatible if they have the same resource type and are joined by a directed path in the sequencing graph.

[2] Equivalently, the definitions can be based on a constraint graph since the vertices are the same in the two models.

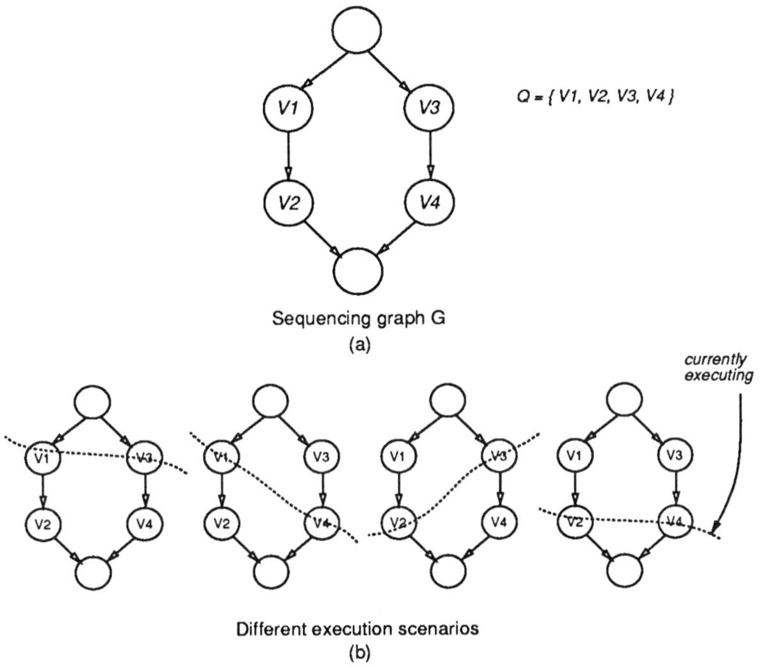

Figure 5.6: Illustrating (a) the sequencing graph G_M with $Q = \{v_1, \ldots, v_4\}$, and (b) different execution scenarios, where the dotted line represents currently executing operations.

Otherwise, computing the concurrency factor based solely on graph connectivity may be too strong, in which case it serves as an upper-bound. Once conflicting vertices have been identified, they are resolved in the conflict resolution phase by serializing the graph subject to execution delays and timing constraints.

We state the following lemma that relates the concurrency factor to G_Q.

Lemma 5.2.1 *Given a sequencing graph G_M without conditional and loop vertices, the concurrency factor of a subset of shareable operations $Q \subseteq \widehat{V}$ is equal to the clique cover number[3] of the corresponding disjoint compatibility graph G_Q.*

[3]The clique cover number is the minimum number of cliques covering a graph [Gol80]. In general, a minimum clique covering is not the same as a minimum clique partition. However, since we are looking for disjoint coverings, coverings and partitions are the same and are used interchangeably in the sequel.

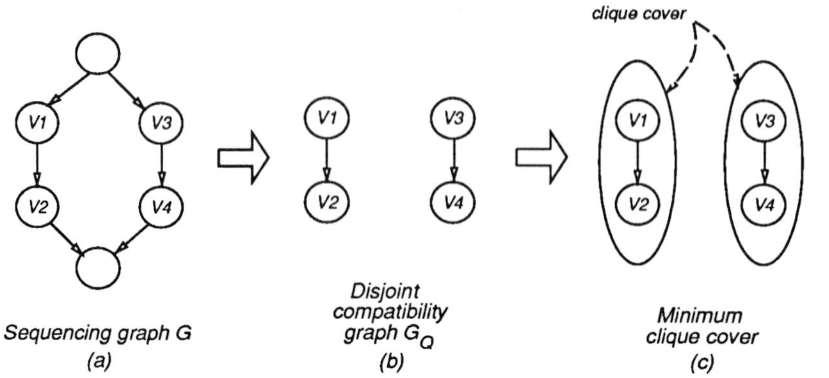

Figure 5.7: Illustrating (a) the sequencing graph G_M with $Q = \{v_1, \ldots, v_4\}$, and (b) the disjoint compatibility graph G_Q of Q. The concurrency factor of Q is the clique cover number of G_Q.

Proof: An edge in the disjoint compatibility graph G_Q implies that two vertices cannot execute in parallel. Therefore, a clique in G_Q corresponds to a subset of vertices that cannot execute in parallel. A clique cover partitions the elements of Q into cliques, and therefore the clique cover number for G_Q is equal to the maximum number of elements in Q that can execute in parallel, which is in turn equal to the concurrency factor of Q. ||

Returning to the example of Figure 5.6, the disjoint compatibility graph G_Q for the set $Q = \{v_1, v_2, v_3, v_4\}$ is shown in Figure 5.7(b). A minimum clique cover is shown in (c) where operations within a clique are enclosed in ovals. The concurrency factor is equal to the clique cover number, which is 2.

Comparability graph.

In general, finding the minimum clique cover for an undirected graph is a NP-complete problem [GJ79]. However, since the disjoint compatibility graph G_Q originates from the sequencing graph model, this worst case complexity can be avoided by exploiting the special properties of G_Q. In particular, there is a class of undirected graphs called *comparability graphs* [Gol80]. A graph $G_Q = (Q, E_Q)$ is called a comparability graph if there exists a *transitive orientation* of G_Q, denoted by (Q, F), such that

$$F \cap F^{-1} = \emptyset, \quad F \cup F^{-1} = E_Q, \quad \mathit{TRANS}(F) \subseteq F$$

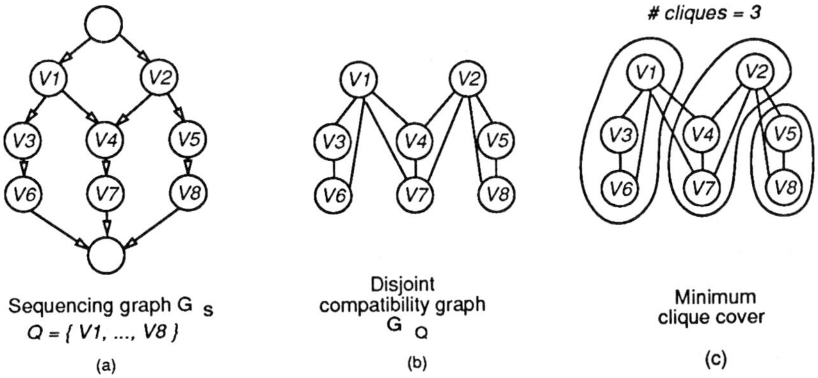

Figure 5.8: Example of computing the concurrency factor by finding clique cover number.

where F^{-1} is the reversal of F, i.e. reversing the direction of all arcs of F, and *TRANS*(F) is the transitive closure of the arcs of F. Another way of describing F is that it is closed under transitivity.

Lemma 5.2.2 *Given a sequencing graph G_M without conditional and loop vertices. The disjoint compatibility graph G_Q for a subset $Q \subseteq V$ is a comparability graph.*

Proof: The transitive orientation F is exactly equal to the transitive closure of the sequencing graph G_M, because the sequencing graph is acyclic and all elements of Q are vertices of G_M. Therefore, G_Q is by definition a comparability graph. ||

The clique cover number for comparability graphs can be computed in polynomial-time using the algorithm presented in Golumbic [Gol80]; we refer the reader to the text for details of the algorithm. Figure 5.8 illustrates the steps in the algorithm applied to a sequencing graph with 8 operations.

5.2.2 Computation for hierarchical graphs

In general, conditionals and loops are present in the model description, and the cf-hierarchy G^*_M consists of multiple sequencing graphs. Accordingly, the elements of $Q \subseteq \widehat{V}$ may reside in different graphs in G^*_M.

To compute the concurrency factor of Q for this general case, we traverse G^*_M in a bottom-up fashion starting from the leaf graphs upwards to the root graph. Let us consider first a leaf graph G_{leaf} with no conditionals or loops. The concurrency factor $cfactor(G_{leaf}, Q)$ can be computed using the strategy presented in the previous section by first constructing a disjoint compatibility graph and then finding its clique cover number.

Now consider the case of a sequencing graph G_i containing conditional and loop vertices. The concurrency factors of Q are assumed to be known for all descendent graphs of G_i in its cf-hierarchy. The strategy to compute the concurrency factor $cfactor(G_i, Q)$ is as follows: a weight $W(v)$ is associated with each vertex $v \in V_i$ of the graph $G_i(V_i, E_i)$ representing the maximum number of elements of Q that can be executed if v is activated. The weight is defined as follows.

- *Loop vertex:* If v is a loop vertex with loop body G_{loop}, then the weight is equal to the concurrency factor of the loop body:

$$W(v) = cfactor(G_{loop}, Q)$$

- *Conditional vertex:* If v is a conditional vertex with k branches $G_{br_1}, \ldots, G_{br_k}$, then the weight is equal to the maximum of the concurrency factors of the branches:

$$W(v) = \max_{i=1}^{k} cfactor(G_{br_i}, Q)$$

- *Elements of Q*: If v is an element of the set Q, i.e. $v \in Q$, then $W(v) = 1$.
- *Otherwise:* If v is none of the above, then $W(v) = 0$.

Consider for example the hierarchy of Figure 5.9 with 4 sequencing graphs. The root graph G_M contains a loop vertex and the loop body contains a conditional vertex with two branches. Q consists of six vertices $\{v_1, \ldots, v_6\}$, where v_1 belongs to the root graph, v_2 belongs to the loop body, v_3 and v_4 belong to branch (cond=1), and v_5 and v_6 belong to branch (cond=2). In Figure 5.9(b), the number in a vertex denotes its weight. The concurrency factors for the two branches, loop body, and root graph are 2, 2, 3, and 4, respectively.

As before, we construct from $G_i(V_i, E_i)$ a disjoint compatibility graph $G_Q = (V_{nz}, E_z)$ induced by the subset of vertices with non-zero weights, denoted by $V_{nz} \subseteq V_i$. From Lemma 5.2.2, G_Q is a comparability graph because a transitive orientation exists for G_Q. If all the weights are 1, then finding a minimum clique cover for G_Q is sufficient to compute the concurrency factor $cfactor(G_i, Q)$.

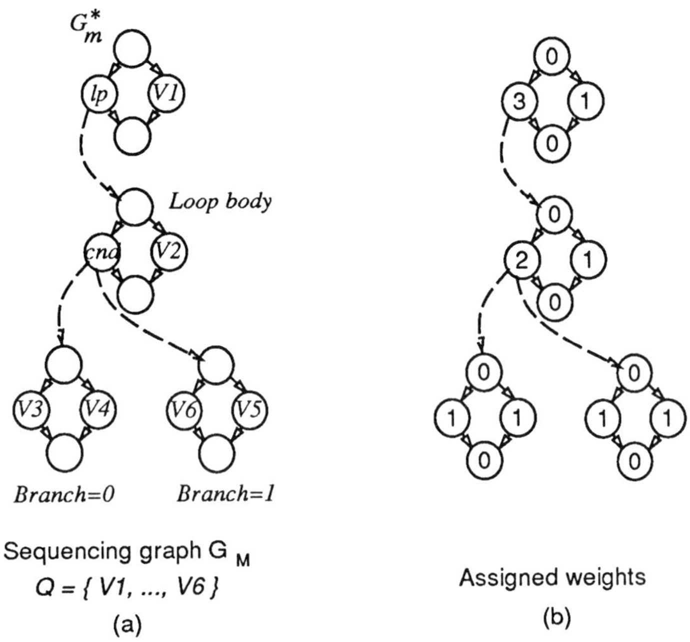

Figure 5.9: Illustrating (a) the cf-hierarchy G^*_M rooted at the sequencing graph G_M. The set Q consists of vertices from the four graphs, $Q = \{v_1, \ldots v_6\}$, (b) corresponding weights.

If not all non-zero weights are equal to 1, we transform G_Q into an *augmented disjoint compatibility graph* $\widetilde{G}_Q = (\widetilde{V}_{nz}, \widetilde{E}_{nz})$. The vertices $\widetilde{V}_{nz}$ are related to V_{nz} via a mapping $\mathcal{A} : V_{nz} \rightarrow \widetilde{V}_{nz}$ that is defined as follows: each vertex $v \in V_{nz}$ is mapped onto one or more vertices $\mathcal{A}(v) = \{\widetilde{v}_{(1)}, \ldots, \widetilde{v}_{(W(v))}\} \subseteq \widetilde{V}_{nz}$ such that the number of target vertices is equal to the vertex weight value, i.e. $|\mathcal{A}(v)| = W(v)$. For example, if $W(v)$ is 1, then one vertex is created in $\widetilde{V}_{nz}$. Alternatively, if $W(v)$ is 5, then five vertices are created in $\widetilde{V}_{nz}$. The edges $\widetilde{E}_{nz}$ are derived as follows: for every edge $(v, w) \in E_{nz}$, we introduce an edge $(p, q) \in \widetilde{E}_{nz}$ for each $p \in \mathcal{A}(v)$ and $q \in \mathcal{A}(w)$. In other words, all vertices in the mapping $\mathcal{A}(v)$ are connected to all vertices in $\mathcal{A}(w)$ in $\widetilde{G}_Q$.

Figure 5.10 illustrates the derivation of an augmented compatibility graph for a graph G_i. The number of vertices in the augmented compatibility graph is equal to the sum of the vertex weights.

Theorem 5.2.1 *Given a sequencing graph G_M and its cf-hierarchy, the concur-*

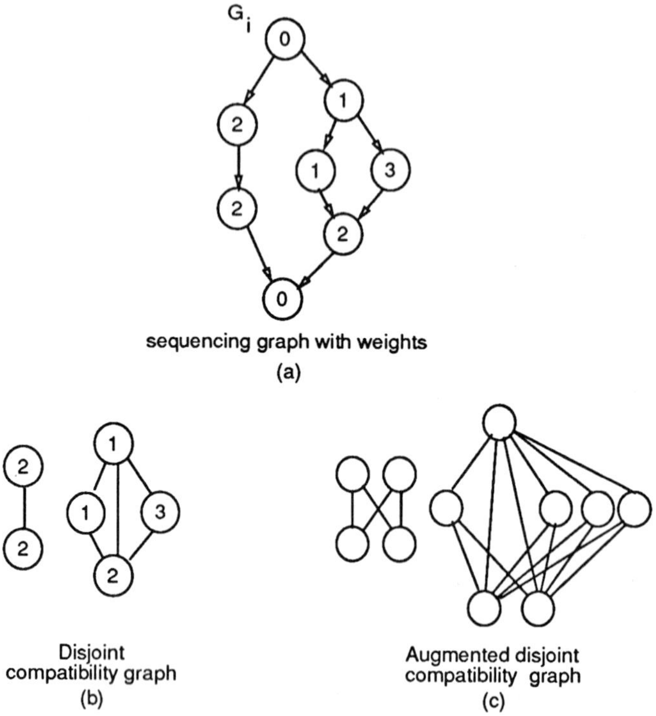

Figure 5.10: Derivation of the augmented disjoint compatibility graph from a graph G_i and appropriate weights: (a) sequencing graph with weights, (b) disjoint compatibility graph, (c) augmented disjoint compatibility graph.

rency factor of a subset of shareable operations $Q \subseteq \hat{V}$ is equal to the clique cover number of the corresponding augmented disjoint compatibility graph $\tilde{G}_Q$.

Proof: The weight for a vertex is equal to the maximum number of vertices in Q that can be activated if the vertex is executed. Consider a vertex v and its mapping $\mathcal{A}(v)$. In the augmented disjoint compatibility graph, no path exists between the elements of $\mathcal{A}(v)$. Furthermore, for any vertex w that is disjoint with respect to v, all elements in the mapping $\mathcal{A}(w)$ are also disjoint w.r.t. all elements in $\mathcal{A}(v)$. Therefore, $\tilde{G}_Q$ captures the degree of parallelism among elements of Q, and we conclude that its clique cover number is equal to the concurrency factor $cfactor(G_M, Q)$. ||

Theorem 5.2.2 *Given a sequencing graph G_M and its cf-hierarchy, the augmented disjoint compatibility graph $\widetilde{G}_Q$ for a subset of shareable operations $Q \subseteq \widehat{V}$ is a comparability graph.*

Proof: Follows directly from the derivation of $\widetilde{G}_Q$ from the corresponding disjoint compatibility graph G_Q. The property of transitivity is not affected, and hence $\widetilde{G}_Q$ is a comparability graph. ||

Since $\widetilde{G}_Q$ is a comparability graph, its clique cover number can be computed in polynomial time [Gol80]. We therefore conclude that given any arbitrary sequencing graph, the concurrency factor for any subset of shareable operations can be computed in polynomial-time.

5.2.3 Conflict-free resource allocation

Concurrency factor can be used to determine the *minimum* resource allocation that is necessary to avoid resource conflicts, where we assume the worst case of all operations having unbounded execution delays. We first consider a sequencing graph G_s and a resource binding β defined on G_s. The resource binding partitions the shareable operations $\widehat{V}$ into one or more instance operation sets $O_{(t,i)}$, where elements within an instance operation set all share the same hardware resource. We define the *conflict degree* of the binding β as follows.

Definition 5.2.2 *The* **conflict degree** *of a resource binding β in a sequencing graph G_s, denoted by $degree(\beta)$, is computed as:*

$$degree(\beta) = \sum_{\forall (t,i) \in \beta} \{cfactor(G_s, O_{(t,i)}) - 1\}$$

The conflict degree is an upper-bound on the number of threads of parallelism that need to be serialized to resolve the resource conflicts in β. Obviously, if $degree(\beta) = 0$, then no resource conflicts exist in the binding. In this case, a logic-level implementation of β exists which satisfies the timing constraints assuming the original timing constraints are valid.

We define the *conflict-free allocation* for a resource type $t \in \mathcal{T}$, denoted by $\alpha_{cf}(t)$, as the minimum allocation of resource type t required to avoid resource conflicts. We state the following theorem.

Theorem 5.2.3 *Given a sequencing graph G_s, the conflict-free allocation of a resource type $t \in \mathcal{T}$ is identical to the concurrency factor of its operation set $O(t)$, i.e. $\alpha_{cf}(t) = cfactor(G_s, O(t))$.*

Proof: We will prove by contradiction. Assume there exists an allocation $\alpha_{lower}(t) \leq \alpha_{cf}(t)$ such that no resource conflicts will arise if $\alpha_{lower}(t)$ resources of type t are allocated. Since $\alpha_{lower}(t) \leq cfactor(G_s, O(t))$, there exists at least one resource binding that is derived from the allocation $\alpha_{lower}(t)$ where two operations bound to the same hardware resource may execute in parallel. This results in a resource conflict and hence contradicts the previous assertion that $\alpha_{lower}(t)$ is a conflict-free allocation of t. Therefore, $\alpha_{cf}(t) = cfactor(G_s, O(t))$ is the conflict-free allocation of t. ||

Corollary 5.2.1 *Given a sequencing graph G_s and a resource allocation $\alpha_{cf}(t)$ for each resource type $t \in \mathcal{T}$, there exists at least one resource binding β_{cf} in the design space such that $degree(\beta_{cf}) = 0$.*

Proof: Follows directly from Theorem 5.2.3. ||

The above theorems imply that conflict-free allocation $\alpha_{cf}(t)$ can be evaluated efficiently because the concurrency factor can be computed in polynomial time. This point is worth emphasizing. The conflict-free allocation indicates the degree of parallelism among the operations of a given resource type. It corresponds to the extent resource conflicts are present in a graph. If the conflict-free allocation satisfies the required resource constraints, then it is not necessary to allocate more resources than this amount to obtain an implementation that satisfies the timing constraints.

On the other hand, if the conflict-free allocation exceeds resource constraints, then additional hardware sharing is required. This results in resource conflicts that must be resolved. Conflict resolution is the subject of Chapter 7.

5.3 Design Space Exploration Strategy

With the design space formulated as a set of resource bindings for a given resource allocation, Hebe explores the design space to find a favorable implementation with respect to a particular design goal, such as minimal area or minimal latency. Any valid implementation must satisfy both resource and timing constraints.

Using the conflict-free allocation as guideline, a set of resource allocations $\{\alpha_1, \ldots, \alpha_k\}$ is specified either by the user manually or by the system automatically[4]. Hebe supports both *exact* and *heuristic* strategies to explore the design space; they are summarized below.

[4]Note that an allocation of $\alpha_i(t)$ means that *exactly* $\alpha_i(t)$ resources are used. Therefore, allocating up to 3 resources is represented by the allocations $\{1, 2, 3\}$.

- *Exact design space exploration:* Exact exploration finds an optimal hardware implementation for a given design. This strategy synthesizes a logic-level implementation for each point in the design space. For many ASIC designs, this is an appropriate strategy because of the restricted size of the design space that stems from the few number of shareable operations and resources.

- *Heuristic design space exploration:* For designs with a large design space, exhaustive synthesis may be prohibitive. To address this difficulty, two heuristic strategies are supported by Hebe. The first strategy constructs only a portion of the design space and the second strategy evaluates and ranks the design space according to a set of cost criteria. The resource bindings with the most favorable cost are synthesized first to determine if they are valid under timing constraints.

The remainder of this chapter describes each strategy in detail.

5.3.1 Exact search strategy

As noted by McFarland [McF87], the area versus performance characteristic for the design space is often highly irregular. Therefore, an optimal solution can be obtained only by synthesizing the entire design space. Exhaustive synthesis is practical for many non-trivial ASIC designs because often a large portion of an ASIC design is application-specific logic functions for which their implementing hardware resources are dedicated. Hebe schedules these operations subject to the timing constraints and generates the corresponding control circuit.

Hebe can compute the *exact* cost for a given design point by performing conflict resolution and scheduling on the design point to generate a logic-level implementation of both control and data-path. This logic-level implementation is passed to logic synthesis for minimization and technology mapping, the result of which provides the exact area and delay cost for the design point.

For a given allocation α, Hebe can generate the complete design space of possible resource bindings. To expedite the search, pruning techniques are applied in traversing both the possible allocations as well as the bindings. Pruning strategies are described below.

- *Pruning resource allocations:* The resource allocations for each resource type $t \in \mathcal{T}$ can be traversed either in decreasing order of allocations $\alpha_i(t)$ if the design goal is minimum latency, or in increasing order if the design goal is minimum area. For the first case, if at any point an allocation produces no valid bindings, then the remaining allocations can be pruned. For the latter case, the first allocation that produces valid bindings is selected. For example, if an allocation of 3 adders and 2 multipliers fails to produce a valid binding, then allocating 2 adders and 1 multipler will also fail to produce a valid binding.

- *Pruning resource bindings:* Hebe must ensure the final implementation satisfies the required timing constraints. One or more bindings may be generated for a given resource allocation α_i. Assuming the original timing constraints are consistent, these generated bindings are *always* valid if there are no maximum timing constraints. However, if maximum timing constraints are specified, then in general not all of these bindings are valid. Conflict resolution is performed to determine whether a binding is valid. Instance operation set(s) which violated the timing constraints are identified and stored in a list. Since any binding having as superset an invalid instance operation set is also invalid, this list can be used to filter the generated bindings.

Pruning ensures that pruned allocations or bindings always violate the required timing constraints. However, many bindings may still result even after pruning. This leads to the heuristic design space exploration strategy, described in the next section.

5.3.2 Heuristic search strategy

For some designs, exact exploration of the design space may be prohibitive due to its size. Hebe also supports heuristic strategies to explore the design space, where the resulting implementation is no longer guaranteed to be optimal.

Two heuristic strategies are used in Hebe: *subspace generation* and *design space ranking*. They are independent strategies that can either be applied individually, or for designs with large design space they can be applied together. For example, subspace generation identifies a portion of the design space as candidate bindings for synthesis. These candidates can be exhaustively synthesized. Alternatively, the entire design space can be considered as possible candidates. These candidates are ranked and synthesized based on the ranking order. For designs with large design space, a portion of the design space is identified as candidates, which are synthesized based on the ranking. The details of each strategy are described in the following sections.

Subspace generation of design space

This section describes a technique to generate a portion of the design space. The points in this subspace correspond to resource bindings that require the *least amount* of serialization to resolve conflicts. The motivation is based on the requirement that the final implementation satisfies the timing constraints. Bindings requiring the least amount of serialization have heuristically better chances of resolving resource conflicts under timing constraints.

The subspace generation strategy is shown in Figure 5.11. It is a divide-and-conquer approach that partitions the operations into *disjoint groups*. Operations in a disjoint group are always implemented by the same hardware resource. The strategy iteratively reduces the number of partition blocks by merging disjoint groups together until the size of the resulting design space, computed using the formula presented in Section 5.1.1, is acceptable to the designer. The major steps in this strategy are described below.

- *Form disjoint groups:* Original dependencies among operations are used to generate an initial partition for each resource type t. The partition is represented as $\{J_1(t), \ldots, J_k(t)\}$. Recall from Section 5.2 that computing the concurrency factor for $O(t)$ involves creating a minimum clique cover, which can be done in polynomial time. Therefore, each clique in this cover is made into a disjoint group because operations within a clique cannot execute in parallel and the number of disjoint groups is the minimum.

- *Compute size of design space:* Given a partition into disjoint groups, we compute the size of the design space if operations within a disjoint group are always bound to the same resource. The size of design space $P(|O(t)|, \alpha(t))$ for a resource type t depends on two factors: the number of operations $|O(t)|$ and the number of allocated resources $\alpha(t)$. By considering only bindings in which all operations in a disjoint group are assigned to the same resource instance, the size of the new design space is equal to $P(cfactor(G_s, O(t)), \alpha(t))$. Since $cfactor(G_s, O(t)) \leq |O(t)|$, the design space size can be decreased.

- *Merge disjoint groups:* If the design space size is acceptable to the designer, then the corresponding resource bindings are generated as candidates for synthesis. Otherwise, the disjoint groups are merged until the design space size is within the limit of a designer specified threshold. The criterion for selecting candidate groups to merge is based on equalizing the

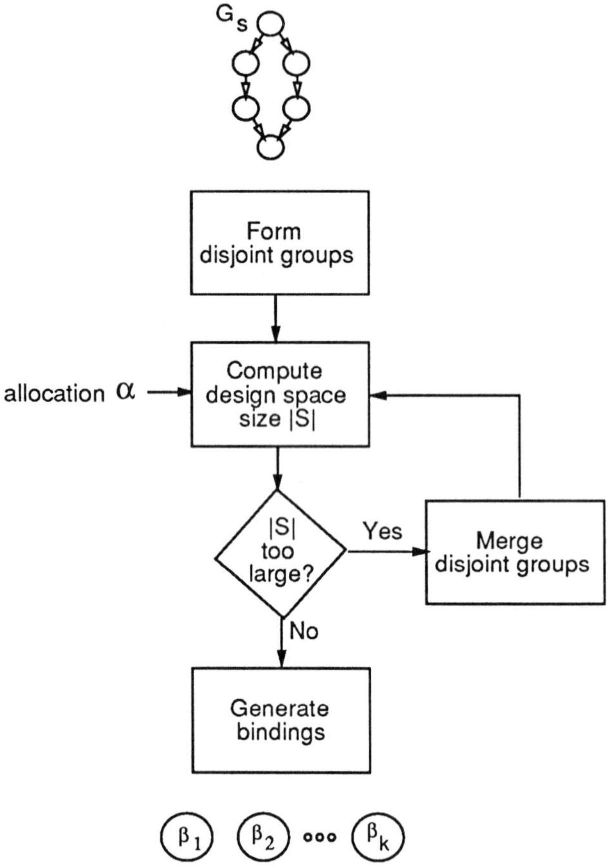

Figure 5.11: Block diagram of subspace generation strategy for heuristic design space exploration.

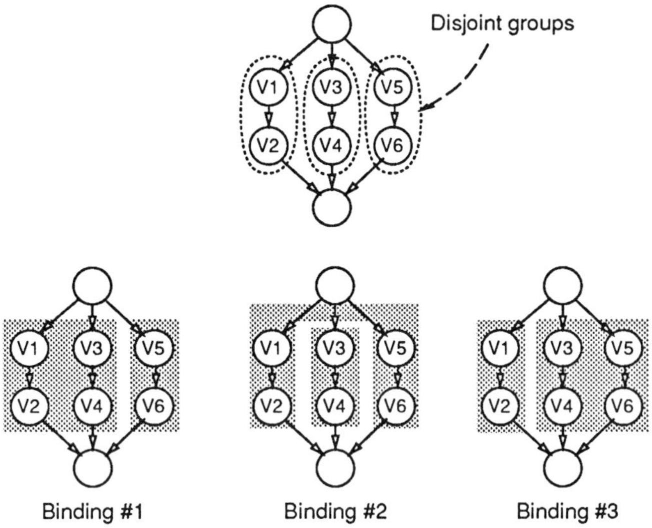

Figure 5.12: Heuristic design space generation for 6 operations to 2 resources: 3 disjoint groups are formed, resulting in three bindings.

sizes of the disjoint groups to minimize the interconnect logic. Each merging increases the concurrency factor of the merged group. The algorithm attempts to minimize the conflict degree of the generated bindings.

To illustrate the approach, consider the sequencing graph in Figure 5.12. There are 6 operations $\{V_1, \ldots, V_6\}$ with the same resource type. Assume 2 resources are allocated. The size of the complete design space is $P(6,2) = 31$. However, by using this generation strategy, 3 disjoint groups are formed: $\{V_1, V_2\}$, $\{V_3, V_4\}$, $\{V_5, V_6\}$. The size of the design subspace is $P(3,2) = 3$, which contains the following bindings: $(1,1,1,1,2,2)$, $(1,1,2,2,1,1)$, $(2,2,1,1,1,1)$.

Ranking the design space

Given a set of candidate resource bindings, the ranking strategy focuses the synthesis effort on the candidates that are most likely to produce good implementations. A block diagram of the ranking strategy is shown in Figure 5.13.

The ranking is based on evaluating the candidate resource bindings with respect to a set of cost criteria. Three cost criteria are supported: *interconnect*,

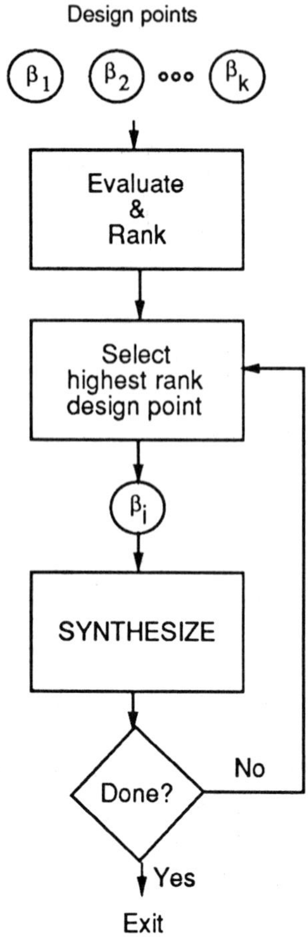

Figure 5.13: Block diagram of the design space ranking strategy in heuristic design space exploration.

area, and *serialization* cost measures, although the framework supports the inclusion of additional cost considerations, such as register count and testability measures. These cost criteria represent estimates of the effect of a particular binding on the area and delay of the final implementation. The bindings with more favorable costs are synthesized first.

The decision of whether one binding is better than another depends on the relative importance of each criteria. This is determined by a *weight* associated with each criterion. The total cost $C_{tot}(\beta)$ for a binding β is given as follows, where C_{int}, C_{area}, and C_{serial} denote the interconnect, area, and serialization costs, respectively.

$$C_{tot}(\beta) = w_{int} \cdot C_{int} + w_{area} \cdot C_{area} + w_{serial} \cdot C_{serial}$$

The values w_{int}, w_{area}, and w_{serial} are weight factors corresponding to each cost criterion. The designer can also specify cut-off *thresholds* for each cost criterion such that a binding with costs exceeding the thresholds will not be considered for synthesis. These three cost criteria are described in detail below.

- *Interconnection cost* C_{int}: The interconnection structure is the steering logic and wiring that guides appropriate values to their proper destinations in the final implementation. Since a resource binding is a complete assignment of operations to resources, the interconnection structure is completely specified. C_{int} is a function of the interconnect's area and delay computed using a multiplexer-based interconnect scheme. More sophisticated area estimates that consider also layout and wiring can also be incorporated at this stage, although they have not been implemented yet in the current version of the system.

- *Area cost* C_{area}: The area cost estimates the total functional area of the final implementation. It considers the area costs due to allocated resources, interconnect structure, registers, and control structure. The major difference between the area and interconnect costs is that C_{int} considers in addition delay characteristics, whereas C_{area} considers only area. C_{int} can be used to obtain more accurate delay estimates prior to scheduling.

- *Serialization cost* C_{serial}: Resource conflicts may arise due to a binding. Determining whether a conflict resolution exists under timing constraints is computationally expensive. Therefore, the *conflict degree* of Section 5.2 is used to estimate the number of threads of parallelism that *need to be serialized* in order to resolve the resource conflicts. C_{serial} is a heuristic measure of the effect of a resource binding on the latency of its implementation. It is defined as $C_{serial} = degree(\beta)$. Obviously, if $C_{serial} = 0$, then no conflict resolution is required, and the implementation for this binding satisfies the required timing constraints.

The candidate resource bindings are ranked according to the weights and pruned according to the thresholds. Synthesis is performed on the candidates with the highest ranking. Through Hebe, the designer can experiment with different design goals by adjusting the values of the weights and focus the synthesis efforts on the resource bindings with acceptable costs.

For example, if the goal is to minimize the area, then area and interconnection costs can be used to identify the resource bindings with minimal area. Likewise, if the goal is to maximize performance under area constraints, then area and interconnection costs can bound the search to those bindings that meet the area constraints, while the serialization cost can provide further pruning of the design space. We emphasize that a resource binding may still be invalid, i.e. violates timing constraints, even when it has favorable costs because conflict resolution and scheduling have not yet been performed.

5.4 Summary of Design Space Exploration

Sharing of hardware resources is important because it characterizes the implementation choices that are possible for a given design. This chapter presented formally the concepts of resource allocation and binding. The design space was formulated as the set of resource bindings for a particular resource allocation. We introduced a closed-form formula to calculate the size of the design space. Concurrency factor was introduced to measure the degree of parallelism among a set of operations. It was then used to derive the conflict-free allocation for a design. We presented approaches to compute the concurrency factors in polynomial-time.

Exact design space exploration exhaustively generates and synthesizes each point in the design space to find an optimal solution. Subspace generation is a heuristic strategy that generates a portion of the design space with the least degree of resource conflicts. Another heuristic strategy is ranking of the design space using cost criteria.

Given a candidate resource binding that has been selected, resource conflicts under timing constraints need to be resolved to determine if the binding is valid. If it is valid, then scheduling is performed and the corresponding control logic is generated. Details of relative scheduling, conflict resolution, and control generation are presented in the remainder of the book.

Chapter 6

Relative Scheduling

As stated earlier, the primary goal of this research is to develop methods of synthesizing hardware from abstract specifications under both *detailed timing* and *synchronization* constraints. Detailed timing constraints capture minimum and maximum bounds on the start time of operations; synchronization constraints model handshaking and coordination among concurrent computation threads, and are represented as operations with *data-dependent* execution delays.

Other existing synthesis approaches either disallow detailed timing constraints by constraining the overall latency only, or they disallow synchronization constraints by assuming all operations to have bounded delays. The remainder of the book describes new formulations and algorithms for high-level synthesis that consider both timing and synchronization constraints. The proposed algorithms are general in nature and can be applied directly to the high-level synthesis framework described in the preceding chapters. In particular, they determine whether a behavior with a particular resource binding is valid, i.e. if a logic-level implementation for this resource binding can be constructed such that timing constraints are satisfied. The behavior is modeled as a constraint graph, which is derived from the sequencing graph with timing constraints; the resource binding is obtained by exploring the design space in Hebe.

The synthesis flow is illustrated in Figure 6.1. First, the original timing and synchronization constraints are analyzed for consistency. If violations occur, then no implementation can be produced to satisfy them and the designer is notified accordingly. Resource conflicts are then checked and resolved, if possible. If resolving conflicts is not possible under the required constraints, then another resource binding is selected from the design space and the process is repeated. If a conflict resolution exists, scheduling is performed to find the start times of the operations. These start times are used to generate a control circuit. Optionally, the resolved graph can be further optimized under timing constraints to reduce the size of the control implementation; this is indicated by the dashed box called *relative control resynchronization*. The data-path is generated by instantiating the allocated functional units, registers, and multiplexer interconnection. Together with the control logic, the final logic-level implementation is described in the SLIF format.

A central idea common to all proposed algorithms is the *relative scheduling* formulation. Relative scheduling provides a framework for analyzing the consistency of a given behavior, modeled by a constraint graph, under both timing and synchronization constraints. In particular, it generalizes the activation of operations with respect to time offsets from the completion of a set of anchors, i.e. data-dependent delay operations. Under this formulation, we can cast resource conflict resolution as the task of serializing operations in the graph that are bound to the same resource component. Analogously, we can cast control optimization as the task of delaying and serializing operations to minimize the size of the control implementation. Timing constraints are used to determine whether a particular serialization is valid or not. Once the graph is appropriately delayed and serialized, relative scheduling is carried out to determine the detailed time offsets that are used to generate the control circuit.

For ease of presentation, relative scheduling is first described in Chapter 6, where resource conflicts are assumed to be resolved. Resource conflict resolution by appropriately serializing the operations subject to the timing constraints is presented in Chapter 7. Control generation from relative scheduling is described in Chapter 8. A novel approach to control optimization called *resynchronization* is described in Chapter 9.

Motivation for relative scheduling. Intuitively, the relative scheduling formulation is an extension of traditional synthesis techniques. Past synthesis algorithms have primarily restricted their scope to operations *within a basic block*, loosely defined as an encapsulation of computations with a single entry point and a single exit point [ASU87]. Once started, all operations within a basic block are executed to completion. The restriction on a single basic block as the target of optimization imposes a overly narrow focus to the synthesis optimizations, and usually produces inferior results for complex designs.

As synthesis research progressed, scheduling and binding techniques evolved to relax the basic block restriction by considering also the *conditional* control-

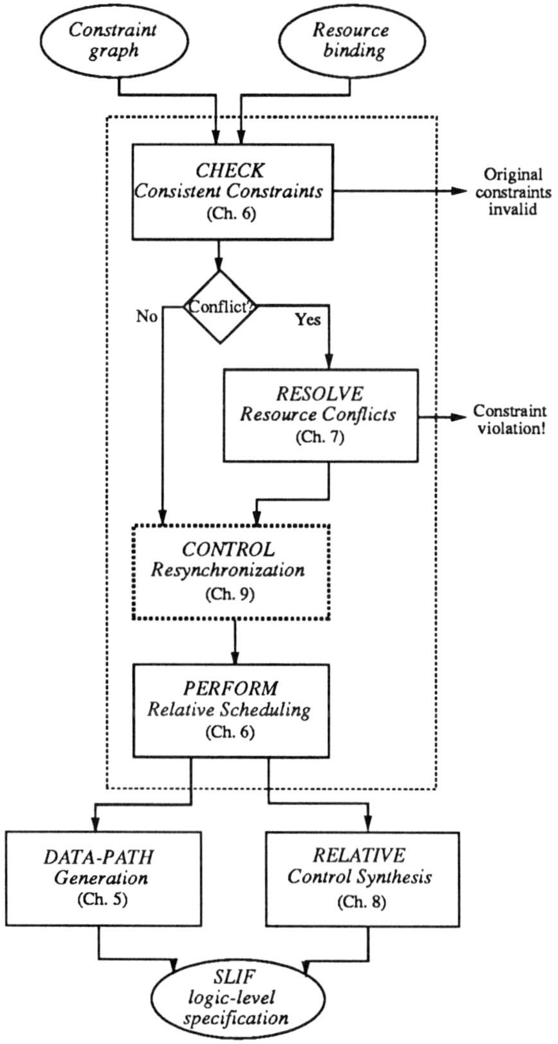

Figure 6.1: Synthesis flow for a selected resource binding.

flow that exists among the basic blocks. Examples of such efforts include Fisher's trace scheduling that considers the global control-flow among basic blocks, Paulin's force-directed scheduling that models conditional execution as probabilistic time intervals, and Wakabayashi's approach that uses conditional vectors to model the mutual exclusivity of operations [CE91]. Although these efforts represent a significant advance in synthesis theory, they still fall short of the ultimate goal of synthesizing large, complex concurrent systems. The requirements of such systems include the ability to reason and manipulate the synchronization among multiple concurrent machines. Even with the support for conditional control-flow, these traditional approaches can only be applied to regions of computations *between* synchronizations. In particular, timing constraints across synchronization domains cannot be rigorously analyzed for consistency.

Relative scheduling addresses this difficulty by proposing a formalism in which points of synchronization are uniformly incorporated as part of the problem formulation. Our approach makes possible global analysis on timing constraints and resource utilization *across* regions of synchronization, thus increasing the scope and applicability of the synthesis algorithms. Furthermore, relative scheduling provides a tigher link to control generation and optimization by more effectively using the global timing and synchronization information to reduce the control implementation cost.

Organization of chapter. This chapter presents the formulation and algorithms for relative scheduling. Our approach can be described in a nutshell as follows. In relative scheduling, we support both operations with fixed delay and operations with data-dependent delay; data-dependent delay operations represent points of synchronization. We uniformly model both types of operations as vertices in the constraint graph model. We assume in this chapter that resource binding and conflict resolution have been performed prior to scheduling.

Consider now the set of operations that depend on a particular synchronization point; we can schedule these operations with respect to the synchronization point under the required timing constraints, as described in Section 6.2. If all operations depend on only a single point of synchronization, then we can apply constraint satisfaction algorithms to find the schedule under timing constraints. In the general case of operations depending on multiple points of synchronization, however, we need to analyze the consistency of timing constraints *across* synchronization points. We introduce in Section 6.2.1 the notion of *well-posed* timing constraints that are satisfied under *any* value of data-dependent delays. We then describe the concept of *redundancy* in Section 6.2.3, where a synchronization point of a particular operation is redundant if it is not needed in computing the operation's start time. Redundancy is used extensively in the scheduling algorithms as well as in the control generation and optimization, described in Chapter 9.

Section 6.3 presents polynomial-time algorithms to check for well-posedness, make the constraints well-posed with minimal serialization, remove redundant anchors, and find the minimum relative schedule. Section 6.4 analyzes the properties of the algorithms. In particular, we show that the algorithms are guaranteed to yield a minimally serialized, well-posed, minimum schedule, if one exists. Finally, Section 6.5 summarizes the relative scheduling approach.

6.1 Introduction to Scheduling

Given a sequencing graph model of hardware behavior and a set of minimum and maximum timing constraints, we derive a *constraint graph* model that is used as the basis for the scheduling formulation and algorithms. The constraint graph model and its derivation are described in Section 4.2. Since the graph model is hierarchical, scheduling is applied hierarchically in a bottom-up fashion. For the sake of simplicity, we consider a non-hierarchical model in this chapter. The extension to hierarchical scheduling is straightforward, and is implemented in Hebe.

For a constraint graph $G(V, E, \omega)$[1], a weight ω_{ij} is associated with each edge (v_i, v_j) that is equal to the execution delay of the operation v_i, denoted by $\delta(v_i)$. Let us assume first that the weights are known; this assumption will be removed in the next section. The scheduling problem may be defined as follows:

Definition 6.1.1 *A* **schedule** *of a constraint graph $G(V, E)$ is an integer labeling $\sigma : V \longrightarrow Z^+$ from the set of vertices V to non-negative integers Z^+, such that $\sigma(v_j) \geq \sigma(v_i) + w_{ij}$ if there is an edge from v_i to v_j with weight w_{ij}. A* **minimum schedule** *is a schedule such that $(\sigma(v_i) - \sigma(v_0))$ is minimum for all $v_i \in V$.*

The integer label $\sigma(v_i)$ associated with a vertex v_i represents the time (or equivalently the cycle) with respect to the beginning of the schedule $(\sigma(v_0))$ in which the operation modeled by v_i may begin execution, i.e. $\sigma(v_i)$ is the *start*

[1]In the sequel, $G(V, E)$ is used to refer to the constraint graph $G(V, E, \omega)$, where ω is made implicit.

time of v_i. The start time of an operation is used by the control to determine when the operation can begin execution.

This scheduling problem, where module binding is performed prior to scheduling, bears similarity to the constrained layout *compaction* problem [LW83] [BN86]. Both problems involve finding the spacing relationships for a set of elements to meet a set of upper and lower bound constraints. In the case of compaction, the elements are objects to be placed on a layout, whereas for scheduling, the elements are operations to be ordered in time. A common goal in both problems is to minimize the total spacing among the elements.

6.2 Relative Scheduling Formulation

Scheduling problems are defined and solved on graphs with fixed delay operations. We extend this notion to graphs with data-dependent delay vertices. For a data-dependent delay vertex v_i, the execution delay $\delta(v_i)$ is not known statically, and can assume any integer value from 0 to ∞. For this reason, we define a subset of the vertices, called *anchors*, that serve as reference points for specifying the start times of operations.

Definition 6.2.1 *The* **anchors** *of a constraint graph $G(V, E)$ consist of the source vertex v_0 and all vertices with data-dependent delay, and are denoted by $A \subseteq V$.*

The source vertex v_0 is treated as an anchor since the activation of a sequencing graph is analogous to the completion of a data-dependent delay source vertex, which is not known a priori. Therefore, all outgoing edges from v_0 have data-dependent weight equal to $\delta(v_0)$.

It is important to point out that the definition of anchors above applies to both *unbounded* delay operations, where the delay is not known statically, and *bounded* delay operations, where the delay is bounded but not fixed, i.e., an operation requiring either 2 or 3 cycles to execute, depending on some condition, is an example of an anchor with bounded delay. It is possible to treat a bounded-delay anchor as a fixed-delay non-anchor by assuming its worst-case delay, at the expense of possibly increasing the overall latency.

We extend the scheduling problem in the presence of data-dependent delay vertices by introducing the concept of *offsets* with respect to the anchors of the graph. Let $V_a \subseteq V$ be the subset of the vertices including a and all its successors. Let $G_a(V_a, E_a)$ be the subgraph induced by V_a, where the execution delays of all data-dependent delay vertices assume the minimum value of zero.

Definition 6.2.2 *The* **offset** *of a vertex* $v_j \in V_a$ *with respect to an anchor* a *is an integer value* $\sigma_a(v_j)$ *such that* $\sigma_a(v_j) \geq \sigma_a(v_i) + w_{ij}$ *if there is an edge of weight* w_{ij} *from* v_i *to* v_j *in* $G_a(V_a, E_a)$, *and* $\sigma_a(a)$ *is normalized to zero. If* $\sigma_a(v_i)$ *is the minimum value, then it is the* **minimum offset** *of* v_i *w.r.t.* a, *and it is denoted by* $\sigma_a^{min}(v_i)$.

Finding the set of offsets is identical to scheduling $G_a(V_a, E_a)$, where the constraint graph models both operation dependencies and timing constraints. If no such set exists, then the constraints are said to be *inconsistent.* Since the execution delay of a data-dependent delay vertex can be any integer greater than or equal to zero, a minimum offset $\sigma_a(v_i)$ is the minimum time after the completion of the anchor a before v_i can begin execution.

We relate now the offsets to the start time of a vertex. Let us consider first the anchors that affect the activation of a vertex v_i.

Definition 6.2.3 *The* **anchor set of a vertex** v_i *is the subset of anchors* $A(v_i) \subseteq A$, *such that* $a \in A(v_i)$ *if there exists a path in* $G_f(V, E_f)$ *from* a *to* v_i *containing at least one data-dependent weight edge with weight equal to* $\delta(a)$.

In other words, an anchor a is in the anchor set of a vertex if the vertex can begin execution only *after* the completion of a. Note that since the graph is polar, the source vertex is contained in the anchor set of every vertex, and the anchor set of the source vertex is the empty set. The anchor set represents the *unknown* factors that affect the activation time of an operation. If we generalize the definition of the *start time* of a vertex in terms of fixed time offsets from the *completion* time of each anchor in its anchor set, then it is possible to completely characterize the temporal relationships among the operations. In particular, the offsets of a vertex can be related to its start time when the execution delays $\{\delta(a), a \in A\}$ of the anchors are known. The **start time** of a vertex v_i, denoted by $T(v_i)$, is defined recursively as follows:

$$T(v_i) \equiv \max_{a \in A(v_i)} \{T(a) + \delta(a) + \sigma_a(v_i)\}$$

Note that if there are no data-dependent delay vertices in the graph, then the start times of all operations will be specified in terms of time offsets from the source vertex, which reduces to the traditional scheduling formulation. We define the relative scheduling problem as follows.

Definition 6.2.4 *A* **relative schedule** Ω *of a constraint graph* $G(V, E)$ *is the set of offsets of each vertex* $v_i \in V$ *with respect to each anchor in its anchor set*

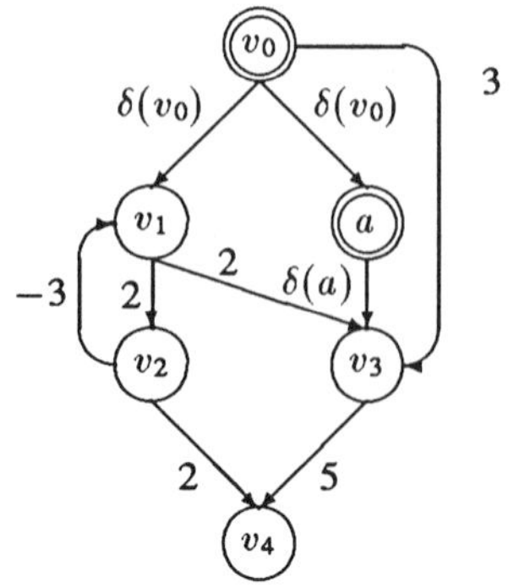

Figure 6.2: Example of a constraint graph, with a maximum timing constraint from v_1 to v_2 and a minimum timing constraint from v_0 to v_3. Vertices v_0 and a are anchors in the graph.

$A(v_i)$, *i.e.* $\Omega = \{\sigma_a(v_i) | a \in A(v_i), \forall v_i \in V\}$. *A* **minimum relative schedule** Ω^{min} *is the set of corresponding minimum offsets, i.e.* $\Omega^{min} = \{\sigma_a^{min}(v_i) | a \in A(v_i), \forall v_i \in V\}$.

A minimum schedule can also be referred to as *as soon as possible* scheduling. A minimum relative schedule for a constraint graph $G(V, E)$ guarantees that, for all profiles of execution delays $\{\delta(a), \forall a \in A\}$, the delay from the source vertex to the sink vertex is minimum. This can easily be shown from the expression for $T(v_i)$ above by noting that if $\sigma_a(v_i)$ is minimum for all v_i, then $T(v_i)$ is also minimum for all v_i. Consider the constraint graph in Figure 6.2. The anchor sets and minimum offsets of the vertices are given in Figure 6.3. For example, vertex v_4 has two anchors v_0 and a with corresponding offsets $\sigma_{v_0} = 8$ and $\sigma_a = 5$; the start time of v_4 is given as:

$$T(v_4) = \max\{T(v_0) + \delta(v_0) + 8, T(a) + \delta(a) + 5\}$$

In words, we say that v_4 begins execution at least 8 cycles after the completion of v_0 *and* at least 5 cycles after the completion of a.

The primary goal of scheduling is to find a schedule that satisfies the required timing constraints. In addition, we identify two secondary goals:

Vertex v_i	Anchor Set $A(v_i)$	Offsets σ_{v_0}	σ_a
v_0	$\emptyset$	-	-
a	$\{v_0\}$	0	-
v_1	$\{v_0\}$	0	-
v_2	$\{v_0\}$	2	-
v_3	$\{v_0, a\}$	3	0
v_4	$\{v_0, a\}$	8	5

Figure 6.3: Illustrating anchor sets and minimum offsets for constraint graph in the previous example.

- Minimize the execution delay of the resulting hardware,
- Minimize the size of the control implementation subject to timing constraints.

These secondary goals may be conflicting, i.e., the schedule that minimizes the latency may be different than the schedule that minimizes the control cost. We address in this chapter the first goal of finding the minimum latency schedule; the second goal is the topic of Chapter 9 on control optimization. Therefore, the term "relative schedule" in the remainder of this chapter refers to the *minimum relative schedule*, unless otherwise stated.

6.2.1 Well-Posedness of timing constraints

An important consideration during scheduling is whether a schedule exists under the required timing constraints. An analysis of the consistency of timing constraints was presented by Camposano and Kunzmann in [CK86] for graphs with no data-dependent delay operations. In this case, a schedule exists if and only if no positive cycles are present in the constraint graph, where a *positive cycle* is a cycle whose sum of the edge weights is a strictly positive integer [LW83]. This condition can be checked by the Bellman-Ford algorithm, or more efficiently, by specialized algorithms [LW83] [BN86].

We extend the analysis in order to consider graphs with data-dependent delay vertices. We first define the notion of *feasible* constraints as follows.

Definition 6.2.5 *A timing constraint is* **feasible** *if it can be satisfied when all data-dependent delays are equal to zero, i.e.* $\delta(a) = 0, \forall a \in A$. *Otherwise, it is* **unfeasible**.

A constraint graph is *feasible* if every constraint in the graph is feasible. Feasibility is a necessary condition for the existence of a schedule. For the special case of no data-dependent delay vertices, the concept of feasibility is also sufficient to ensure that a schedule for the constraint graph exists. We state the necessary and sufficient condition for feasible constraints in the following theorem.

Theorem 6.2.1 *A constraint graph* $G(V, E)$ *is feasible if and only if no positive cycle exists in* G, *assuming data-dependent delays in* G *are set to zero.*

Proof: Let $G_0(V, E)$ denote the constraint graph $G(V, E)$ where all the data-dependent delays are set to zero. We prove first the necessary condition. If the constraint graph G is feasible, then all constraints in G_0 must be consistent. Let $\Omega = \{\sigma(v_i) | \forall v_i \in V\}$ denote a schedule of the constraint graph $G_0(V, E)$ satisfying the constraints. Consider now a cycle in the graph, denoted by $(v_1, v_2), (v_2, v_3), \cdots, (v_{s-1}, v_s), (v_s, v_1)$. The inequality constraints implied by the edges of the cycle are as follows:

$$\begin{aligned}
\sigma(v_1) + w_{1,2} &\leq \sigma(v_2) \\
\sigma(v_2) + w_{2,3} &\leq \sigma(v_3) \\
&\cdots \\
\sigma(v_{s-1}) + w_{s-1,s} &\leq \sigma(v_s) \\
\sigma(v_s) + w_{s,1} &\leq \sigma(v_1)
\end{aligned}$$

Adding the inequalities above, we have

$$\sigma(v_1) + (\text{sum of edge weights on cycle}) \leq \sigma(v_1)$$

Since all constraints are consistent, the above inequality must also be satisfied. Therefore, the length of the cycle must not be positive. This is true for any cycle in the graph G_0, and we conclude that no positive cycle exists in the graph.

Conversely, assume that no positive cycles exist in the graph. Then we define $\sigma^{LP}(v)$ to be the length of the longest path from the source vertex v_0 to v assuming the data-dependent delays are set to zero, i.e. $\sigma^{LP}(v) = length(v_0, v)$. We will show that $\{\sigma^{LP}(v), v \in V\}$ is a solution set, i.e. for any edge $e_{ij} \in E$ with weight w_{ij}, the inequality $\sigma^{LP}(v_i) + w_{ij} \leq \sigma^{LP}(v_j)$ is satisfied, which will imply the constraint graph is feasible.

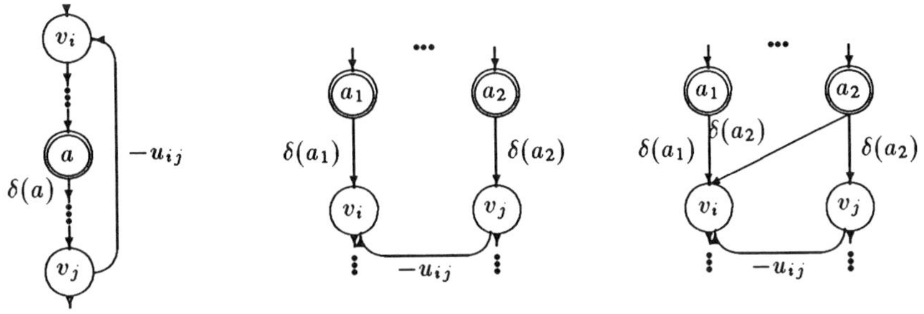

Figure 6.4: Examples of ill-posed timing constraints (a) and (b), and well-posed constraint (c), where the double-circled vertices are anchors with data-dependent delays.

Assume for the sake of contradiction that there exists an edge $e_{ij} \in E$ such that the constraint is violated, i.e.

$$\sigma^{LP}(v_i) + w_{ij} > \sigma^{LP}(v_j)$$

The above inequality implies the longest path from v_0 to v_i does not pass through the edge e_{ij}, since otherwise $\sigma^{LP}(v_i)+w_{ij}$ would be equal to $\sigma^{LP}(v_j)$. Specifically, the path from v_0 to v_j consisting of the longest path from v_0 to v_i, followed by the edge e_{ij}, is longer than the longest path from v_0 to v_j. This contradicts the definition of $\sigma^{LP}(v_j)$, and hence the above inequality is not true. Since the previous argument holds for all edges, $\{\sigma^{LP}(v), v \in V\}$ is a solution set and the graph is feasible. ||

We now consider the consistency of constraints in the presence of data-dependent delay vertices. Intuitively, the data-dependent delay vertices create time gaps that cannot be resolved statically. Depending on the execution profile of these operations, a timing constraint *may* or *may not* be satisfied by a given schedule. We extend the analysis by introducing the concept of *well-posed* versus *ill-posed* timing constraints, in the presence of data-dependent delay operations.

Definition 6.2.6 *A timing constraint is* **well-posed** *if it can be satisfied for all values of execution delays of the data-dependent delay vertices.*

Conversely a timing constraint is said to be *ill-posed* if it cannot be satisfied for some values of the data-dependent delays. A constraint graph $G(V, E)$ is well-posed if every constraint implied by the edges E is well-posed. From the definition of feasible constraints, if a graph is well-posed, then it is also necessarily feasible. The contra-positive also holds; specifically, if a graph is unfeasible, then it is ill-posed. Because of the observation that no schedule exists for unfeasible constraint graphs, we assume in subsequent analysis the constraint graphs to be feasible, unless otherwise indicated.

Note that minimum timing constraints are always feasible and well-posed, because the check for their validity does not depend on the values of the execution delays, as explained in the previous section. On the other hand, a maximum timing constraint defines an upper-bound between the activation of two operations. If its satisfiability depends on the completion time of a data-dependent delay vertex, then the constraint cannot be met in general because it is possible that an input data sequence exists such that the execution delay of the data-dependent delay vertex exceeds the upper-bound imposed by the constraint.

Consider the examples in Figure 6.4. Both graphs contain an *ill-posed* maximum timing constraint u_{ij} from v_i to v_j, represented by a backward edge (v_j, v_i) with weight $-u_{ij}$. In Figure 6.4(a), a data-dependent delay vertex a exists on the path from v_i to v_j. Depending on how long it takes to complete execution, the constraint may or may not be satisfied. Similarly for Figure 6.4(b), the activation of v_i depends on the completion of a_1, and the activation of v_j depends on the completion of a_2, both of which are data-dependent.

Assume for the sake of simplicity that the anchor set of v_i consists of a_1 and the source vertex v_0, and the anchor set of v_j consists of a_2 and the source vertex v_0. The start times for v_i and v_j can be written as:

$$\begin{aligned} T(v_i) &= \max\{T(a_1) + \delta(a_1) + \sigma_{a_1}(v_i), T(v_0) + \delta(v_0) + \sigma_{v_0}(v_i)\} \\ T(v_j) &= \max\{T(a_2) + \delta(a_2) + \sigma_{a_2}(v_j), T(v_0) + \delta(v_0) + \sigma_{v_0}(v_j)\} \end{aligned}$$

Since $T(v_i)$ does not depend on a_2, and $T(v_j)$ does not depend on a_1, the satisfiability of a maximum timing constraint between v_i and v_j depends on the data-dependent delays $\delta(a_1)$ and $\delta(a_2)$, making it ill-posed.

Consider, however, the situation in Figure 6.4(b) if we introduce a forward edge from a_2 to v_i with data-dependent edge weight equal to $\delta(a_2)$, as shown in Figure 6.4(c). In this case the constraint will become well-posed. The reason is because by the time v_i begins execution (after the completion of both a_1 and a_2), all the data-dependent delays in the fan-in of v_j is already known, i.e. $\delta(a_2)$ is common to both $T(v_i)$ and $T(v_j)$. The satisfiability of the constraint can therefore be determined independently of data-dependent delays. We formalize this observation in stating the following lemma as a necessary and sufficient condition for checking if a given maximum timing constraint is well-posed.

Lemma 6.2.1 *Let $G(V, E_f)$ be acyclic. A feasible maximum timing constraint $u_{ij} \geq 0$ is well-posed if and only if $A(v_j) \subseteq A(v_i)$.*

Proof: We prove first the necessary condition. For the sake of contradiction, assume $A(v_j)$ is not a subset of $A(v_i)$ and the maximum timing constraint u_{ij} is well-posed. The start time of v_j is $T(v_j) = \max_{a \in A(v_j)} \{T(a) + \delta(a) + \sigma_a(v_j)\}$, and the start time of v_i is $T(v_i) = \max_{a \in A(v_i)} \{T(a) + \delta(a) + \sigma_a(v_i)\}$. The maximum timing constraint implies the condition $T(v_j) \leq T(v_i) + u_{ij}$. The inequality can be written as:

$$\begin{aligned} T(v_j) - T(v_i) &\leq u_{ij} \\ \max_{a \in A(v_j)} \{T(a) + \delta(a) + \sigma_a(v_j)\} - \max_{a \in A(v_i)} \{T(a) + \delta(a) + \sigma_a(v_i)\} &\leq u_{ij} \end{aligned}$$

Since $A(v_j)$ is not a subset of $A(v_i)$, there exists an anchor b such that $b \in A(v_j)$ but $b \notin A(v_i)$. Thus it is always possible to find a value of $\delta(b)$ such that the inequality is violated. Hence, the constraint graph is ill-posed.

We now prove the sufficient condition. If the anchor sets of v_i and v_j for a feasible maximum timing constraint u_{ij} satisfy the condition $A(v_j) \subseteq A(v_i)$, then the constraint implies the following inequality:

$$\begin{aligned} T(v_j) &\leq T(v_i) + u_{ij} \\ \max_{a \in A(v_j)} \{T(a) + \delta(a) + \sigma_a(v_j)\} &\leq \max_{a \in A(v_i)} \{T(a) + \delta(a) + \sigma_a(v_i)\} + u_{ij} \\ \max_{a \in A(v_j)} \{T(a) + \delta(a) + \sigma_a(v_j)\} &\leq \max_{a \in A(v_i)} \{T(a) + \delta(a) + (\sigma_a(v_i) + u_{ij})\} \\ \max_{a \in A(v_j)} \{T(a) + \delta(a) + \sigma_a(v_j)\} &\leq \max\{ \max_{a \in A(v_j)} \{T(a) + \delta(a) + (\sigma_a(v_i) + u_{ij})\} \\ &\qquad \max_{x \in A(v_i), x \notin A(v_j)} \{T(x) + \delta(x) + (\sigma_x(v_i) + u_{ij})\}\} \\ \max_{a \in A(v_j)} \{T(a) + \delta(a) + \sigma_a(v_j)\} &\leq \max\{\mathcal{A}, \mathcal{B}\} \end{aligned}$$

where we define $\mathcal{A} \equiv \max_{a \in A(v_j)} \{T(a) + \delta(a) + (\sigma_a(v_i) + u_{ij})\}$, and $\mathcal{B} \equiv \max_{x \in A(v_i), x \notin A(v_j)} \{T(x) + \delta(x) + (\sigma_x(v_i) + u_{ij})\}$. It is sufficient to verify that $\max_{a \in A(v_j)} \{T(a) + \delta(a) + \sigma_a(v_j)\} \leq \mathcal{A}$, because $\max_{a \in A(v_j)} \{T(a) + \delta(a) + \sigma_a(v_j)\} \leq \mathcal{A}$ implies $\max_{a \in A(v_j)} \{T(a) + \delta(a) + \sigma_a(v_j)\} \leq \max\{\mathcal{A}, \mathcal{B}\}$. Therefore,

$$\max_{a \in A(v_j)} \{T(a) + \delta(a) + \sigma_a(v_j)\} \leq \mathcal{A}$$
$$\max_{a \in A(v_j)} \{T(a) + \delta(a) + \sigma_a(v_j)\} \leq \max_{a \in A(v_j)} \{T(a) + \delta(a) + (\sigma_a(v_i) + u_{ij})\}$$

Note that all quantities in the inequality above are non-negative. Since both the left hand and right hand side of the inequality refer to the same set of anchors, determining whether it can be satisfied can be stated in terms of the individual anchors. In particular, for all anchors $a \in A(v_j)$, the following inequality is checked.

$$T(a) + \delta(a) + \sigma_a(v_j) \leq T(a) + \delta(a) + (\sigma_a(v_i) + u_{ij})$$
$$\sigma_a(v_j) \leq \sigma_a(v_i) + u_{ij}$$

By the definition of feasible timing constraints, the inequality holds for all offsets $\sigma_a(v_i), v_i \in V$ and anchors $a \in A(v_j)$. Therefore, the maximum timing constraint u_{ij} is satisfied. ||.

Lemma 6.2.2 *Given a well-posed constraint graph $G(V, E)$, the anchor sets of the vertices on a cycle of G are identical.*

Proof: Let a cycle be formed in the graph by the edges (v_1, v_2) (v_2, v_3) $\cdots$ (v_{s-1}, v_s) (v_s, v_1). The edges can be classified either as *forward* or *backward*. We consider each case separately. If (v_{k-1}, v_k) is a backward edge, then $A(v_{k-1}) \subseteq A(v_k)$ by Lemma 6.2.1 because of the well-posedness property. If (v_{k-1}, v_k) is a forward edge, then from the definition of anchor sets, $A(v_{k-1}) \subseteq A(v_k)$ because v_{k-1} is the predecessor of v_k. Combining the two requirements, the edges in the cycle imply that,

$$A(v_1) \subseteq A(v_2) \subseteq \cdots \subseteq A(v_s) \subseteq A(v_1)$$

which can be true if and only if the anchor sets are identical,

$$A(v_1) \equiv A(v_1) \equiv \cdots \equiv A(v_s)$$

for all cycles in the graph. ||

Since two cycles in the graph with a common vertex is also a cycle, the anchor sets for the vertices on all connected cycles are identical. A direct corollary of the lemma is the following.

Corollary 6.2.1 *Given a well-posed constraint graph $G(V, E)$, no cycles with data-dependent length exist in G.*

Proof: We will prove by contradiction. Assume G is well-posed but there exists a cycle with data-dependent length. Let the cycle be denoted by $\mathcal{C}$. Since $\mathcal{C}$ has data-dependent length, this implies that there exists an anchor a on the cycle such that the length of the cycle is greater than or equal to the execution delay $\delta(a)$. Consider now the next vertex v that follows a on the cycle $\mathcal{C}$. By definition of anchor sets, a is in the anchor set of v, i.e. $a \in A(v)$. From Lemma 6.2.2, the anchor sets of all vertices on the cycle must be identical, implying that a is also in the anchor set of a itself. This results in a contradiction. Therefore, we conclude that no cycle of data-dependent length exists in G. ||

With Lemma 6.2.1, and Lemma 6.2.2, we state the following key theorem.

Theorem 6.2.2 *Let $G(V, E_f)$ be acyclic. A feasible constraint graph $G(V, E)$ is well-posed if and only if $A(v_i) \subseteq A(v_j)$ for all edges $e_{ij} \in E$.*

Proof: First we prove the necessary condition by induction. We will show that a given well-posed constraint graph, if an edge e_{ij} is added such that $A(v_i) \subseteq A(v_j)$, then the resulting graph is well-posed also. Initially, consider the graph consisting of forward edges E_f only. Since $G(V, E_f)$ is acyclic and by the definition of anchor sets the condition holds, and $G(V, E_f)$ is well-posed. Now consider a backward edge $e_{ij} \in E_b$ representing a feasible maximum timing constraint u_{ji}, where by assumption $A(v_i) \subseteq A(v_j)$. From Lemma 6.2.1, u_{ji} is well-posed if and only if $A(v_i) \subseteq A(v_j)$. Therefore, the resulting graph is well-posed also, and the induction is complete.

Now we prove the sufficient condition. Assume $G(V, E)$ is well-posed and there exist an edge $e_{ij} \in E$ for which $A(v_i)$ is not a subset of $A(v_j)$. By definition of anchor sets, e_{ij} cannot be a forward edge, and hence e_{ij} must be a backward edge that is derived from a feasible maximum timing constraint. Since all constraints implied by G are well-posed, it follows from Lemma 6.2.1 that $A(v_i) \subseteq A(v_j)$. This results in a contradiction. Therefore, the criterion $A(v_i) \subseteq A(v_j)$ must be satisfied for all edges in the graph. ||

6.2.2 Properties of relative schedule

In this section, we analyze several properties of relative scheduling. The following theorem states the existence criterion for making a constraint graph well-posed.

Lemma 6.2.3 *A feasible constraint graph $G(V, E)$ can be made well-posed if and only if no data-dependent length cycles exist in G.*

Proof: We prove first the sufficient condition. If no data-dependent length cycle exists, we prove by induction that it is possible to satisfy the well-posedness condition for all edges. As the basis of the induction, consider the forward constraint graph G_f. By definition of anchor sets, G_f is well-posed. Now consider a backward edge $e_{ij} \in E_b$. If $A(v_i) \subseteq A(v_j)$, the constraint is well-posed. Otherwise, there exists an anchor $x \in A(v_i)$ but $x \notin A(v_j)$. By assumption, there are no data-dependent length cycles. Therefore there must not be a path from v_j to x since otherwise the data-dependent cycle (from $x \stackrel{\delta(x)}{\rightarrow} v_i \stackrel{e_{ij}}{\rightarrow} v_j \rightarrow x$) would be formed. Because of this observation, we can add an edge from x to v_j without creating an data-dependent length cycle. This can be done for all $\{x | x \in A(v_i), x \notin A(v_j)\}$, and e_{ij} can be made well-posed without creating an data-dependent length cycle. The induction is complete and a well-posed solution exists by considering all backward edges until all edges are well-posed.

Now we prove the necessary condition by showing that if it is possible to make a constraint graph well-posed, then no data-dependent length cycles exist. Assume G can be made well-posed by addition of a set of edges E_{add}, such that $\widetilde{G}(V, E \cup E_{add})$ is well-posed. From Corollary 6.2.1, there are no data-dependent length cycles in $\widetilde{G}$. The introduction of additional edges E_{add} does not affect the original cycles in G, i.e. any cycle in the original graph will remain a cycle in the final graph by the addition of E_{add}. Therefore, if $\widetilde{G}$ has no data-dependent length cycles, then no data-dependent length cycle can exist in G also. The proof is complete. ||

Given the existence criterion for well-posed constraints, we state the following theorem that interprets the *minimum relative schedule* $\Omega^{min} = \{\sigma_a^{min}(v_i) | a \in A(v_i), \forall v_i \in V\}$ in terms of the lengths of the longest paths in the constraint graph. Assume $G(V, E)$ to be well-posed, this implies that there are no positive cycles in the graph (by feasibility). Let $\Omega^{LP} = \{\sigma_a^{LP}(v_i) | a \in A(v_i), \forall v_i \in V\}$ be the relative schedule where the offset $\sigma_a^{LP}(v_i)$ w.r.t. the an anchor $a \in A(v_i)$ is the length of the longest path from a to v_i in the constraint graph $G(V, E)$, or equivalently, $\sigma_a^{LP}(v_i) = length(a, v_i)$. We show now the equivalence $\Omega^{min} \equiv \Omega^{LP}$.

Theorem 6.2.3 *Assume the constraint graph $G(V, E)$ to be well-posed. Then for all offsets $\sigma_a^{LP}(v_i) \in \Omega^{LP}$ and $\sigma_a^{min}(v_i) \in \Omega^{min}$, $\sigma_a^{LP}(v_i) = \sigma_a^{min}(v_i)$.*

Proof: The proof uses an extension of the analysis presented in [LW83]. We show first that $\Omega^{LP} = \{\sigma_a^{LP}(v_i) | a \in A(v_i), \forall v_i \in V\}$ is a relative schedule of $G(V, E)$. For the sake of contradiction, assume there exists an edge $e_{ij} \in G$ with weight w_{ij} that does not satisfy the inequality constraint. This implies that there exists an anchor a common to both anchor sets, $a \in A(v_i) \cap A(v_j)$, such that the corresponding offsets violate the inequality constraint. Let $\sigma_a^{LP}(v_i)$ and $\sigma_a^{LP}(v_j)$ denote the offset of v_i and v_j w.r.t. anchor a, respectively. Then the violation is given as

$$\sigma_a^{LP}(v_i) + w_{ij} > \sigma_a^{LP}(v_j)$$

The inequality implies that the longest weighted path from the anchor a to v_j does not contain the edge e_{ij}, otherwise, $\sigma_a^{LP}(v_i) + w_{ij}$ would be equal to $\sigma_a^{LP}(v_j)$. If we consider the path from a-to-v_j as consisting of the longest weighted path from a-to-v_i and the edge e_{ij}, then the sum of edge weights on this path is greater than $\sigma_a^{LP}(v_j)$. This is contrary to the definition of the longest path from a-to-v_j, and the inequality cannot be true. Since the inequality holds for all anchors common to $A(v_i)$ and $A(v_j)$ for all edges e_{ij} in the graph, the set Ω_{LP} is a relative schedule of G.

We still need to prove that $\Omega^{LP} = \{\sigma_a^{LP}(v_i) | a \in A(v_i), \forall v_i \in V\}$ is the *minimum* relative schedule. Let $\Omega' = \{\sigma'_a(v_i) | a \in A(v_i), \forall v_i \in V\}$ be any relative schedule satisfying the inequalities implied by G. We need to show that for all offsets $\sigma_a^{LP}(v_i) \in \Omega_{LP}$ and $\sigma'_a(v_i) \in \Omega'$, $\sigma_a^{LP}(v_i) \leq \sigma'_a(v_i)$ for all anchors $a \in A(v_i)$ of all vertices $v_i \in V$. For an anchor $a \in A(v_i)$, the offset $\sigma_a^{LP}(v_i)$ is defined to be the length of the longest weighted path from a to v_i. Represent this path as $(a, v_i^1)(v_i^1, v_i^2) \cdots (v_i^{s-1}, v_i^s)$, where $v_i^s = v_i$. Let $W(v_i, v_j)$ denote the weight associated with the edge (v_i, v_j), then $\sigma_a^{LP}(v_i) = \sum_{k=1}^{s} W(v_i^{k-1}, v_i^k)$. The anchor sets $A(v_i^k)$, $1 \leq k \leq s$ on the path above all contain the anchor a because it is the predecessor of every vertex v_i^k, $1 \leq k \leq s$. The relative schedule $\Omega'(v_i)$ satisfies all the following inequalities,

$$\begin{aligned} \sigma'_a(a) + W(a, v_i^1) &\leq \sigma'_a(v_i^1) \\ \sigma'_a(v_i^1) + W(v_i^1, v_i^2) &\leq \sigma'_a(v_i^2) \\ &\cdots \\ \sigma'_a(v_i^{s-1}) + W(v_i^{s-1}, v_i^s) &\leq \sigma'_a(v_i) \end{aligned}$$

where $\sigma'_a(v_i^k)$ is the offset of v_i^k w.r.t. the anchor a in Ω'. Adding the inequalities above, we have

$$\sigma'_a(a) + \textstyle\sum_{k=1}^{s} W(v_i^{k-1}, v_i^k) \leq \sigma'_a(v_i)$$

We normalized the offset $\sigma'_a(a)$ to zero, and hence by definition,

$$\sigma_a^{LP}(v_i) = \sum_{k=1}^{s} W(v_i^{k-1}, v_i^k) \leq \sigma'_a(v_i)$$

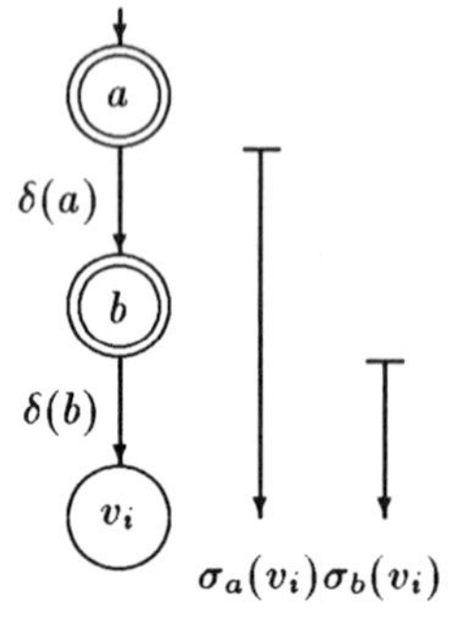

Figure 6.5: Illustrating the cascading effect of the anchor, where a and b are anchors of v_i.

for all anchors $a \in A(v_i)$. This implies $\sigma_a^{LP}(v_i) \leq \sigma'_a(v_i)$ for all $a \in A(v_i)$ and $v_i \in V$. We conclude that Ω^{LP} is the minimum relative schedule of G. ||

6.2.3 Relevant and irredundant anchor sets

An important issue that arises in the calculation of start times is the *cascading* effect of anchors. Let $A(v_i)$ and $T(v_i)$ be the anchor set and start time of a vertex v_i, where $T(v_i)$ is given as:

$$T(v_i) \equiv \max_{a \in A(v_i)} \{T(a) + \delta(a) + \sigma_a(v_i)\}$$

In general, there may be disjoint paths to a vertex v_i from every anchor $a \in A(v_i)$. Therefore, the corresponding offsets $\sigma_a(v_i), \forall a \in A(v_i)$ are necessary in the computation of the start time $T(v_i)$.

However, consider the case of a path of anchors followed by the vertex v_i, as shown in Figure 6.5. Note that $A(v_i)$ includes both anchors a and b. Since the anchor b can begin execution only after a completes, and since v_i can begin execution only after b completes, it is sufficient to define $T(v_i)$ with respect to the completion of b *only*. In other words, anchor a is dominated by b, and hence $\sigma_a(v_i)$ is not needed to compute $T(v_i)$.

We formalize this observation by identifying the anchors in the anchor set whose offsets are not necessary in computing the start time. These anchors and the corresponding offsets can be removed without altering the result of subsequent scheduling steps. The advantages of removing redundancies are two-fold.

First, we improve significantly the efficiency of the scheduling algorithm (Section 6.3.5) by focusing on a smaller number of anchors. Second, we can achieve a smaller and faster control implementation of a relative schedule because the start time depends on fewer offsets, and hence on fewer synchronizations.

This section is organized as follows. We introduce first the concept of the *relevant anchor set* of a vertex as points of reference that may *directly* affect its start time. We demonstrate in Theorem 6.2.4 that for the case of well-posed constraints and minimum offsets, the start time computed using the relevant anchor set is equivalent to the start time computed using the full anchor set. We then use the relevant anchors as the basis for defining *irredundant anchors.* We present Theorem 6.2.6 which states that the irredundant anchors of a vertex are the *minimum* set of anchors necessary to compute its start time. The theorem sets the theoretical framework for our use of irredundant anchors in the scheduling algorithm.

Relevant anchor set

We identify the anchors of a vertex v_i that may *directly* affect the start time $T(v_i)$ by introducing the concept of the *relevant anchor set* of a vertex. We present first the following definitions.

Definition 6.2.7 *A* **defining path** *of an anchor $a \in A$ to a vertex v_i, denoted by $\rho(a, v_i)$, is a path from a-to-v_i in $G(V, E)$ wth exactly one data-dependent edge such that the weight on the data-dependent edge is equal to $\delta(a)$. The length of $\rho(a, v_i)$, denoted by $|\rho(a, v_i)|$, is the sum of the edge weights on the path excluding the data-dependent weight $\delta(a)$.*

Definition 6.2.8 *The* **relevant anchor set of a vertex** *v_i in a constraint graph $G(V, E)$ is the set of anchors $R(v_i) = \{r \mid r \in A$ such that there exists a defining path $\rho(r, v_i)\}$.*

Definition 6.2.9 *A* **maximal defining path** *of a relevant anchor $r \in R(v_i)$ of a vertex v_i is the defining path $\rho^*(r, v_i)$ whose length is maximal among all defining paths from r-to-v_i.*

Therefore, the maximal defining path of a relevant anchor $r \in R(v_i)$ is the longest path by which r is defined to be relevant. Obviously, there can be more than one defining path for a particular relevant anchor since there can be more than one path from r to v_i. Similarly, the maximal defining paths may not be unique.

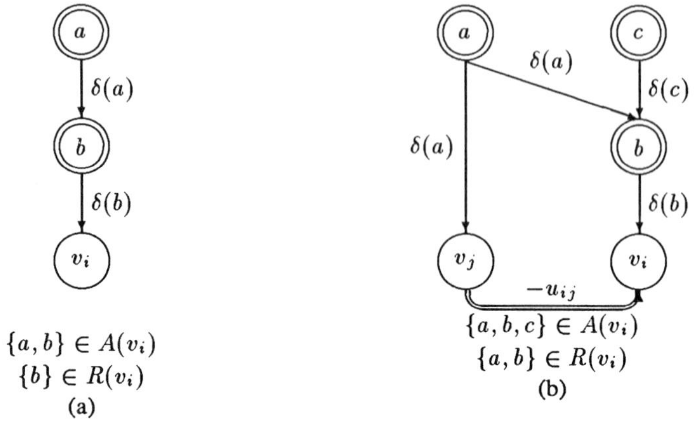

Figure 6.6: Illustrating the difference between the full anchor set $A(v_i)$ and the relevant anchor set $R(v_i)$: (a) b is a relevant anchor of v_i, (b) both a and b are relevant anchors of v_i.

It is important to point out that the definition of relevant anchor set considers all paths in the *full* constraint graph, as opposed to the definition of anchor set that considers only paths in the forward constraint graph. This is because the definition of the full anchor sets is independent of the property of well-posedness, whereas the relationship between the relevant anchor set with the full anchor set strongly depends on the property of well-posedness. To illustrate the concept, consider the examples in Figure 6.6. The double-line edge from v_j to v_i is a backward edge, all other edges are forward edges. For Figure 6.6(b), a is an relevant anchor of v_i because there is a bounded path from a-to-v_j-to-v_i, which is the defining path of $a \in R(v_i)$.

We define the **relevant start time** of a vertex v_i, denoted by $T_R(v_i)$, as the start time of v_i computed with offsets from the relevant anchors $R(v_i)$ only. Specifically,

$$T_R(v_i) \equiv \max_{r \in R(v_i)} \{T(r) + \delta(r) + \sigma_r(v_i)\}$$

We present now several properties of relevant anchor sets that are important in proving the equivalence between the start time and the relevant start time of a vertex. First, the following lemma casts the property of well-posedness in terms of the relationship between the relevant anchor set and the anchor set.

Lemma 6.2.4 *Let $G(V, E_f)$ be acyclic. A feasible constraint graph $G(V, E)$ is well-posed if and only if $R(v_i) \subseteq A(v_i)$ for all $v_i \in V$.*

Proof: From Theorem 6.2.2, it is sufficient to show that $R(v_i) \subseteq A(v_i)$ $\forall v_i \in V$ implies $A(v_i) \subseteq A(v_j)$ $\forall e_{ij} \in E$, and vice versa. We prove the sufficient condition first. Assume that $A(v_i) \subseteq A(v_j)$ for all $e_{ij} \in E$. Consider a relevant anchor $r \in R(v_i)$ of a vertex $v_i \in V$. By definition, there exists a defining path from r to v_i, denoted by $(r, v_1), (v_1, v_2), \cdots, (v_k, v_i)$, such that r is an anchor and $\{v_1, \cdots, v_k\}$ are not anchors. By assumption of well-posedness, the edges imply that

$$A(v_1) \subseteq A(v_2) \subseteq \cdots \subseteq A(v_k) \subseteq A(v_i)$$

Since $r \in A(v_1)$, this implies that $r \in A(v_i)$ for all relevant anchors $r \in R(v_i)$. Therefore, $R(v_i) \subseteq A(v_i)$.

We prove the necessary condition by contradiction. Assume $R(v_i) \subseteq A(v_i)$ for all $v_i \in V$. Assume also there exists vertices v_i and v_j such that $e'_{ij} \in E$, and the condition $A(v_i) \subseteq A(v_j)$ is violated. Then there exists an anchor x such that $x \in A(v_i)$ and $x \notin A(v_j)$. By the definition of anchor sets, the violating edge e'_{ij} cannot be a forward edge. Since $x \in A(v_i)$, there is a path of forward edges from x to v_i with the data-dependent edge weight $\delta(x)$. If the path contains no other data-dependent delay edges, then we have a contradiction because x is a relevant anchor of v_j (by the defining path from x to v_i followed by the edge e'_{ij}), but $x \notin A(v_j)$.

Now consider the case when the path contains other data-dependent delay edges. In particular, let an anchor q be on the path from x to v_i such that $q \in A(v_i)$. However, q cannot be in the anchor set $A(v_j)$, otherwise we would violate our initial assumption of $x \notin A(v_j)$. By replacing x by q, the same argument can be applied. Therefore, the condition $A(v_i) \subseteq A(v_j)$, $\forall e_{ij} \in E$ is satisfied, and the graph is well-posed. ||

For ease of notation, let $X(v_i) = \{x | x \in A(v_i), x \notin R(v_i)\}$ denote the set of *irrelevant anchors* of v_i, where $A(v_i) = R(v_i) \cup X(v_i)$. The following lemma states the relationships between the relevant and irrelevant anchors of a vertex.

Lemma 6.2.5 *Consider a well-posed constraint graph $G(V, E)$. For each irrelevant anchor $x \in X(v_i)$, x is a predecessor of at least one relevant anchor $r \in R(v_i)$.*

Proof: Consider an anchor $x \in A(v_i)$, then there is a path of forward edges from x-to-v_i. Denote the path as $(x, v_1)(v_1, v_2) \cdots (v_k, v_i)$. Assume $x \notin R(v_i)$, there must exist by definition at least one anchor on the path. Let the last anchor on the path be denoted as v_a, where $v_a \in A(v_i)$. This implies that v_a is a relevant anchor of v_i, i.e. $v_a \in R(v_i)$. Since there is a path of forward edges from x-to-v_a, x is a predecessor of v_a. The argument holds for all paths from x to v_i. Therefore, x is the predecessor of at least one relevant anchor of v_i. ||

In other words, the set of relevant anchors of a vertex v_i forms a *vertex separation-set* on the subgraph induced by all the paths to v_i from every anchor $a \in A(v_i)$. We can now present the following theorem that demonstrates the equivalence between the start time and the relevant start time, in the presence of well-posed constraints and minimum offsets.

Theorem 6.2.4 *Let $G(V, E)$ be a well-posed constraint graph with a minimum relative schedule $\Omega^{min} = \{\sigma_a^{min}(v_i) | a \in A(v_i), \forall v_i \in V\}$. Then the corresponding start time $T(v_i)$ is equivalent to the relevant start time $T_R(v_i)$ for all $v_i \in V$.*

Proof: We will prove by induction. We note that since the forward constraint graph $G(V, E_f)$ is acyclic, there exists a topological ordering $<$ of the vertices such that $v_i < v_j$ if v_i is a predecessor of v_j. Consider each vertex according to its topological ordering, starting with the source vertex v_0. Obviously $T(v_0) = T_R(v_0)$ since $A(v_0) = R(v_0) = \emptyset$. Now consider the next vertex v_1, since the graph is polar, the equality is again satisfied because the anchor set and the relevant anchor set are identical and equal to the source vertex v_0, i.e. $A(v_1) = R(v_1) = \{v_0\}$.

The inductive hypothesis assumes that for a vertex v_i, if all vertices $v_0, \cdots, v_{i-1}$ that precede it in the ordering satisfies the equality in start times, then the equality will also be satisfied for v_i. Note this implies that for all anchor $a \in A(v_i)$, $T(a) = T_R(a)$ because the anchors of v_i are also predecessors of v_i, and hence precede v_i in the topological ordering.

We will now show that $T(v_i) = T_R(v_i)$. Expanding the expressions for $T(v_i)$ and $T_R(v_i)$,

$$\begin{aligned} T(v_i) &= \max_{a \in A(v_i)} \{T(a) + \delta(a) + \sigma_a^{min}(v_i)\} \\ &= \max\{ \max_{r \in R(v_i)} \{T(r) + \delta(r) + \sigma_r^{min}(v_i)\}, \\ &\qquad \max_{x \in X(v_i))} \{T(x) + \delta(x) + \sigma_x^{min}(v_i)\}\} \end{aligned}$$

$$T_R(v_i) = \max_{r \in R(v_i)} \{T(r) + \delta(r) + \sigma_r^{min}(v_i)\}$$

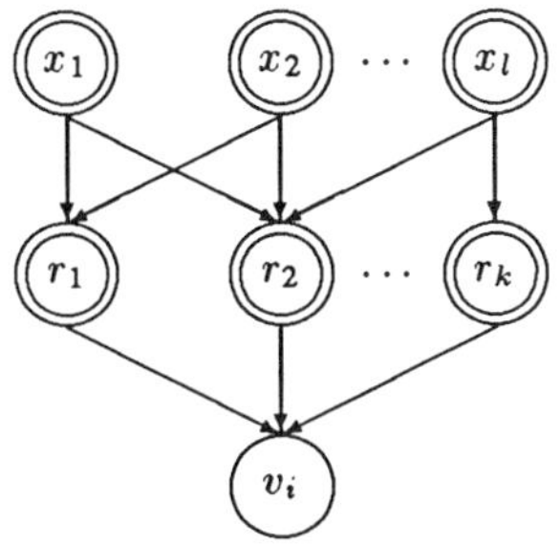

Figure 6.7: Illustrating the relevant and irrelevant anchors of v_i, where all edges represent paths in the graph with data-dependent length.

If the following inequality is satisfied, then the equality $T_R(v_i) = T(v_i)$ is satisfied:

$$\max_{r \in R(v_i)} \{T(r) + \delta(r) + \sigma_r^{min}(v_i)\} \geq \max_{x \in X(v_i)} \{T(x) + \delta(x) + \sigma_x^{min}(v_i)\}$$

where $X(v_i)$ is the set of irrelevant anchors of v_i. Let $X(v_i) = \{x_1, \cdots, x_l\}$ and let $R(v_i) = \{r_1, \cdots, r_k\}$. We expand the inequality above as follows:

$$\begin{array}{rcl} \mathcal{L} = \max\{T(r_1) + \delta(r_1) + \sigma_{r_1}^{min}(v_i), & & \mathcal{R} = \max\{T(x_1) + \delta(x_1) + \sigma_{x_1}^{min}(v_i), \\ T(r_2) + \delta(r_2) + \sigma_{r_2}^{min}(v_i), & \geq & T(x_2) + \delta(x_2) + \sigma_{x_2}^{min}(v_i), \\ \cdots & & \cdots \\ T(r_k) + \delta(r_k) + \sigma_{r_k}^{min}(v_i)\} & & T(x_l) + \delta(x_l) + \sigma_{x_l}^{min}(v_i)\} \end{array}$$

where for notational convenience we define $\mathcal{L}$ to be the left hand side of the inequality, and $\mathcal{R}$ the right hand side. By Lemma 6.2.5, any $x \in X(v_i)$ must be a predecessor of at least one relevant anchor $r \in R(v_i)$. Figure 6.7 shows the relevant anchors $R(v_i)$ forming a vertex separation-set between the irrelevant anchors $X(v_i)$ and v_i in the subgraph induced by all the paths to v_i from $A(v_i)$. Since the irrelevant anchors $x_1, \cdots, x_l$ are predecessors of the relevant anchors $r_1, \cdots, r_k$, we can rewrite the start time of the relevant anchors $T(r_1), \cdots, T(r_k)$ in terms of the irrelevant anchors. In particular,

$$
\begin{aligned}
T(r_1) &= \max_{x^1 \in A(r_1)} \{T(x_1^1) + \delta(x_1^1) + \sigma_{x_1^1}^{min}(r_1), \\
&\qquad T(x_2^1) + \delta(x_2^1) + \sigma_{x_2^1}^{min}(r_1), \\
&\qquad \cdots\} \\
T(r_2) &= \max_{x^2 \in A(r_2)} \{T(x_1^2) + \delta(x_1^2) + \sigma_{x_1^2}^{min}(r_2), \\
&\qquad T(x_2^2) + \delta(x_2^2) + \sigma_{x_2^2}^{min}(r_2), \\
&\qquad \cdots\} \\
&\cdots \\
T(r_k) &= \max_{x^k \in A(r_k)} \{T(x_1^k) + \delta(x_1^k) + \sigma_{x_1^k}^{min}(r_k), \\
&\qquad T(x_2^k) + \delta(x_2^k) + \sigma_{x_2^k}^{min}(r_k), \\
&\qquad \cdots\}
\end{aligned}
$$

where $\{x_1^1, x_2^1, \cdots\}$ is the set of anchors for the relevant anchor r_1, and likewise for $\{x_1^2, x_2^2, \cdots\}$, and so on. We know that the set of irrelevant anchors $\{x_1, \cdots, x_l\}$ are equal to the union of the anchor sets of the relevant anchors, i.e. $X(v_i) = \bigcup_{j=1}^{k} A(r_j)$. This follows from Lemma 6.2.5 because the relevant anchors $R(v_i)$ form a vertex separation set between the irrelevant anchors $X(v_i)$ and v_i. For each of the terms in $\mathcal{L}$, we replace $T(r_i)$ by the corresponding expression, then rearrange the terms to reexpress in terms of the set of irrelevant anchors $\{x_1, \cdots, x_l\}$. The expression $\mathcal{L}$ can be rewritten as follows.

$$
\begin{aligned}
\mathcal{L} = \max\{T(x_1) + \delta(x_1) &+ \max_{r^1 \in \Gamma(x_1)} \{\sigma_{x_1}^{min}(r^1) + \delta(r^1) + \sigma_{r^1}^{min}(v_i)\}, \\
T(x_2) + \delta(x_2) &+ \max_{r^2 \in \Gamma(x_2)} \{\sigma_{x_2}^{min}(r^2) + \delta(r^2) + \sigma_{r^2}^{min}(v_i)\}, \\
&\cdots \\
T(x_l) + \delta(x_l) &+ \max_{r^l \in \Gamma(x_l)} \{\sigma_{x_l}^{min}(r^l) + \delta(r^l) + \sigma_{r^l}^{min}(v_i)\}\}
\end{aligned}
$$

Note that we have regrouped the expressions in terms of the irrelevant anchors $\{x_1, \cdots, x_l\}$. The set $\Gamma(x_1)$ consists of the relevant anchors of v_i which lie on all paths from x_1 to v_i in the full graph $G(V, E)$. Similarly, the set $\Gamma(x_2)$ consists of the relevant anchors of v_i which lie on all paths from x_2 to v_i in $G(V, E)$, and so on. Comparing the expression above with the preceding inequality, we note that a sufficient condition to satisfy the preceding inequality ($\mathcal{L} \geq \mathcal{R}$) is if the following new set of inequalities are satisfied:

$$\begin{aligned}
\max_{r^1 \in \Gamma(x_1)} \{\sigma_{x_1}^{min}(r^1) + \delta(r^1) + \sigma_{r^1}^{min}(v_i)\} &\geq \sigma_{x_1}^{min}(v_i) \\
\max_{r^2 \in \Gamma(x_2)} \{\sigma_{x_2}^{min}(r^2) + \delta(r^2) + \sigma_{r^2}^{min}(v_i)\} &\geq \sigma_{x_2}^{min}(v_i) \\
&\dots \\
\max_{r^l \in \Gamma(x_l)} \{\sigma_{x_l}^{min}(r^l) + \delta(r^l) + \sigma_{r^l}^{min}(v_i)\} &\geq \sigma_{x_l}^{min}(v_i)
\end{aligned}$$

Let us consider the first inequality in the set above. Any path from x_1 to v_i contains at least one relevant anchor $r^1 \in \Gamma(x_1)$. From Theorem 6.2.3, the minimum offsets $\{\sigma_a^{min} | a \in A(v_i), \forall v_i \in V\}$ for a constraint graph $G(V, E)$ correspond to the lengths of longest paths from the anchors to their successors, where all data-dependent edge weights are set to zero. Since all paths from x_1 to v_i contain at least one relevant anchor, by the assumption of minimum offsets, which are equal to the lengths of the longest paths (Theorem 6.2.3), the following equality holds:

$$\sigma_{x_1}^{min}(v_i) = \quad length(x, v_i) = \quad \max_{r^1 \in \Gamma(x_1)} \{\sigma_{x_1}^{min}(r^1) + \sigma_{r^1}^{min}(v_i)\}$$

Since $\delta(r^1) \geq 0, \ \forall r^1 \in \Gamma(x_1)$, the first inequality is satisfied. A similar argument can be applied to the other inequalities. We conclude then the start times of v_i is equal to its relevant start time, $T(v_i) \equiv T_R(v_i)$, and the induction is complete. ||

Irredundant anchor set

A relevant anchor *may* directly affect the activation of a vertex. However, redundancies could still arise. Consider the example in Figure 6.8, where both a and b are relevant anchors of v_i. Then a is *redundant* in the computation of the start time of v_i since there is a path (through a-b-v_i) with length at least as long as the length of the maximal defining path $\rho^*(a, v_i)$ of a (through a-v_1-v_i). We generalize the above observation by defining the notion of *redundancy* in the relevant anchor set.

Definition 6.2.10 *An anchor $r \in A(v_i)$ of a vertex v_i is* **redundant** *if there exists an anchor q such that:*

1. *$r \in A(q)$ and $q \in A(v_i)$, and*

2. *$length(r, v_i) = length(r, q) + length(q, v_i)$.*

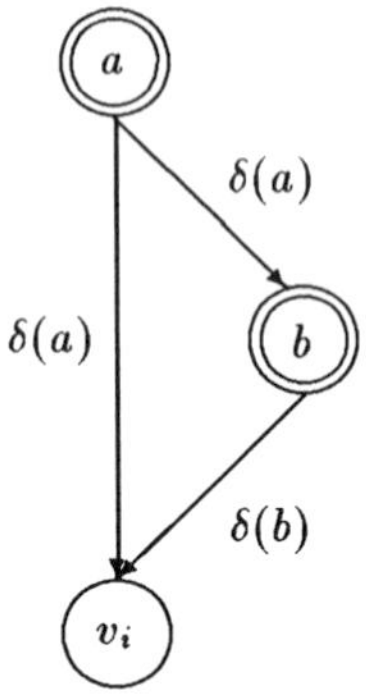

Figure 6.8: Example of a redundant anchor a of the vertex v_i.

Otherwise, r is an **irredundant** *anchor of v_i. The set of irredundant anchors of v_i is denoted by $IR(v_i)$.*

Theorem 6.2.5 *An irredundant anchor $a \in IR(v_i)$ of a vertex v_i is always a relevant anchor of v_i, i.e. $IR(v_i) \subseteq R(v_i)$.*

Proof: Consider an irrelevant anchor $x \in X(v_i)$. By Lemma 6.2.5, the relevant anchors form a vertex separation set between x and v_i, such that for all paths ρ from x to v_i, ρ contains a relevant anchor $r \in R(v_i)$. Therefore, $length(x, v_i) = \max_{q \in \Gamma(x)} \{length(x, q) + length(q, v_i)\}$. This implies that there exists a relevant anchor q such that $length(x, v_i) = length(x, q) + length(q, v_i)$. This is exactly the definition of redundancy. Since all irrelevant anchors are redundant, all irredundant anchors must be relevant, i.e. $IR(v_i) \subseteq R(v_i)$ for all $v_i \in V$. ||

To illustrate the concept, consider the two graphs in Figure 6.9. In (a), a is irredundant since there is a maximal defining path of a (through a-v_1-v_3) that is the longest path from a to v_3, assuming data-dependent weights are set to zero. In (b), a is redundant because the length of its maximal defining path is less than $length(a, v_3)$.

The above theorem states that $IR(v_i) \subseteq R(v_i)$. We present now Lemma 6.2.6 which states that the start time computed using $IR(v_i)$ only, denoted by $T_{IR}(v_i) = \max_{r \in IR(v_i)} \{T(r) + \delta(r) + \sigma_r^{min}(v_i)\}$, is *equivalent* to $T(v_i)$ computed using the full anchor set $A(v_i)$, for well-posed constraints and minimum offsets. $T_{IR}(v_i)$ is called the *irredundant start time* of v_i.

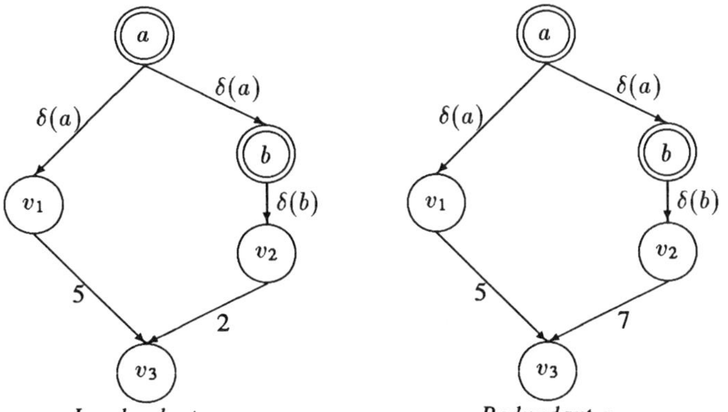

Figure 6.9: Illustrating the difference between irredundant and redundant anchors.

Lemma 6.2.6 *Let $G(V, E)$ be a well-posed constraint graph with a minimum relative schedule $\Omega^{min} = \{\sigma_a^{min}(v_i) | a \in A(v_i), \forall v_i \in V\}$. The corresponding start time $T(v_i)$ is equivalent to the irredundant start time $T_{IR}(v_i)$ for all $v_i \in V$.*

Proof: By Theorem 6.2.4, $T(v_i) \equiv T_R(v_i)$. It is sufficient therefore to show that $T_{IR}(v_i) \equiv T_R(v_i), \forall v_i \in V$. We will prove by induction in a similar manner as in Theorem 6.2.4. We note that since the forward constraint graph $G(V, E_f)$ is acyclic, there exists a topological ordering $<$ of the vertices such that $v_i < v_j$ if v_i is a predecessor of v_j. Consider each vertex according to its topological ordering, starting with the source vertex v_0. Obviously $T_R(v_0) = T_{IR}(v_0)$ since $A(v_0) = R(v_0) = IR(v_0) = \emptyset$. Now consider the next vertex v_1, since the graph is polar, the equality is again satisfied because the relevant anchor set and irredundant anchor set are identical and equal to the source vertex v_0, i.e. $R(v_1) = IR(v_1) = \{v_0\}$.

The inductive hypothesis assumes that for a vertex v_i, if all vertices $v_0, \cdots, v_{i-1}$ that precede it in the ordering satisfies the equality in start times, then the equality will also be satisfied for v_i. This implies that for all anchor $a \in A(v_i)$, $T_R(a) = T_{IR}(a)$ because the anchors of v_i are predecessors of v_i, and hence precede v_i in the topological ordering. We will now show that $T_R(v_i) = T_{IR}(v_i)$. We expand the expression for the start times as follows.

$$\begin{aligned} T_R(v_i) &= \max_{r \in R(v_i)} \{T(r) + \delta(r) + \sigma_r^{min}(v_i)\} \\ &= \max\{ \max_{r \in IR(v_i)} \{T(r) + \delta(r) + \sigma_r^{min}(v_i)\}, \\ &\qquad \max_{x \notin IR(v_i)} \{T(x) + \delta(x) + \sigma_x^{min}(v_i)\}\} \end{aligned}$$

$$T_{IR}(v_i) = \max_{r \in IR(v_i)} \{T(r) + \delta(r) + \sigma_r^{min}(v_i)\}$$

If the following inequality is satisfied, then $T_R(v_i) = T_{IR}(v_i)$.

$$\max_{r \in IR(v_i)} \{T(r) + \delta(r) + \sigma_r^{min}(v_i)\} \geq \max_{x \notin IR(v_i)} \{T(x) + \delta(x) + \sigma_x^{min}(v_i)\}$$

Let us consider one redundant relevant anchor $x \notin IR(v_i)$ but $x \in R(v_i)$. For all maximal defining path of x to v_i, there exists a path from x-to-v_i passing through at least one irredundant anchor q with longer length, since otherwise x is irredundant. By definition, there exists a path κ from x to v_i in $G(V, E)$ containing an irredundant relevant anchor $q \in IR(v_i)$, such that the length of κ is equal to the length of the longest weighted path from x-to-v_i, i.e. $|\kappa| = length(x, v_i)$. By Theorem 6.2.3, $|\kappa|$ is equal to $\sigma_x^{min}(v_i)$, which is greater than or equal to the length of any maximal defining path of x, i.e. $|\kappa| \geq |\rho^*(x, v_i)|, \forall \rho^*(x, v_i)$. Furthermore, since the longest path contains anchor q, $\sigma_x^{min}(v_i)$ is equal to the sum of the offsets $\sigma_x^{min}(q) + \sigma_q^{min}(v_i)$. The relationship is summarized as follows:

$$\begin{aligned} T(x) + \delta(x) + \sigma_x^{min}(v_i) &\leq T(q) + \delta(q) + \sigma_q^{min}(v_i) \\ T(x) + \delta(x) + \sigma_x^{min}(v_i) &\leq [T(x) + \delta(x) + \sigma_x^{min}(q)] + \delta(q) + \sigma_q^{min}(v_i) \\ \sigma_x^{min}(v_i) &\leq \sigma_x^{min}(q) + \delta(q) + \sigma_q^{min}(v_i) \end{aligned}$$

For every redundant anchor $x \notin IR(v_i)$, there exist an irredundant anchor $q \in IR(v_i)$ that dominates the offset $\sigma_x^{min}(v_i)$. Since $\delta(q) \geq 0$, the inequality above is satisfied, and the induction is complete. ||

With Lemma 6.2.6, we state the following important theorem.

Theorem 6.2.6 *Let $G(V, E)$ be a well-posed constraint graph with a minimum relative schedule $\Omega^{min} = \{\sigma_a^{min}(v_i) | a \in A(v_i), \forall v_i \in V\}$, The irredundant anchor set $IR(v_i)$ is the* **minimum** *set of anchors that is required to compute the start time $T(v_i), \forall v_i \in V$.*

Proof: The sufficient condition is already satisfied by Lemma 6.2.6. We prove the necessary condition by showing that if an irredundant anchor $r \in$

$IR(v_i)$ is not used, then the resulting start time $\overline{T}^{min}(v_i)$ will violate one or more constraints implied by the edges of $G(V, E)$.

By Theorem 6.2.5, the irredundant anchors are also relevant anchors. Therefore, there exists a maximal defining path $\rho^*(r, v_i)$ of $r \in IR(v_i)$, where the $|\rho^*(r, v_i)|$ is equal to $length(r, v_i)$. By Theorem 6.2.3, $length(r, v_i)$ is equal to the minimum offset $\sigma_r^{min}(v_i)$. Let $T^{min}(v_i)$ and $T^{min}(r)$ be the start times of v_i and r computed with the minimum offsets, where the data-dependent edge weights are set to their minimum value of 0. Since $T^{min}(v_i)$ and $T^{min}(r)$ satisfy the constraints of $G(V, E)$, the following equality must be satisfied:

$$\begin{aligned} T^{min}(v_i) &= T^{min}(r) + length(r, v_i) \\ T^{min}(v_i) &= T^{min}(r) + \sigma_r^{min}(v_i) \end{aligned}$$

Because r is irredundant, for all paths κ from r to v_i containing one or more anchors $\{a_1, \cdots, a_k\}$ such that $r \in A(a_1), a_1 \in A(a_2), \cdots, a_k \in A(v_i)$, the length of κ is less than the longest path from r to v_i, i.e. the following condition holds:

$$\sigma_r^{min}(v_i) \quad > \quad \sigma_r^{min}(a_1) + \sigma_{a_1}^{min}(a_2) + \cdots + \sigma_{a_k}^{min}(v_i)$$

If r is not included in the computation of the start time of v_i, then the offset $\sigma_r^{min}(v_i)$ is not included in the expression for $T^{min}(v_i)$. Let $\overline{T}^{min}(v_i)$ denote the start time of v_i computed without the offset $\sigma_r^{min}(v_i)$, where all data-dependent edge weights are set to zero. Let $|\kappa| = \sigma_r^{min}(a_1) + \sigma_{a_1}^{min}(a_2) + \cdots + \sigma_{a_k}^{min}(v_i)$ denote the length of the longest path from r to v_i excluding the maximal defining paths from r-to-v_i. Then the equality satisfied by the new start time $\overline{T}^{min}(v_i)$ is:

$$\overline{T}^{min}(v_i) = T^{min}(r) + |\kappa|$$

However, since $|\kappa| < \sigma_r^{min} = length(r, v_i)$, the inequality implied by the longest path from r to v_i in the constraint graph is violated, i.e. $\overline{T}^{min}(v_i) < T^{min}(v_i)$. We conclude that r must be used to compute the start time of v_i. The same argument applies to every irredundant anchor, and hence $IR(v_i)$ is necessary to compute the start time of v_i. ||

6.3 Algorithms for Relative Scheduling

Given a sequencing graph and a set of minimum and maximum timing constraints, we generate a constraint graph $G(V, E)$ consisting of forward edges E_f

and backward edges E_b. The forward edges are first checked to ensure that no cycles are formed. We approach the relative scheduling problem in four steps, as shown in Figure 6.10.

1. *Checking Well-posed* – The constraint graph is checked for well-posedness using Theorem 6.2.2, using an algorithm called *checkWellposed.*

2. *Making Well-posed* – If the constraint graph is ill-posed, then no schedule can satisfy the constraints for all input sequences. We can however attempt to make it well-posed by adding sequencing dependencies to selectively serialize the graph. This is performed by the algorithm *makeWellposed,* which is guaranteed to yield a well-posed graph with *minimum* serialization, if one exists. If the constraint graph cannot be made well-posed, then the set of constraints is inconsistent, and we exit the algorithm.

3. *Remove Redundant Anchors* – At this point, the constraint graph is well-posed. We then identify and remove the redundant anchors that are not needed to compute the start times.

4. *Iterative Incremental Scheduling* – Finally, the relative schedule can be computed by decomposing the constraint graph into a set of subgraphs for each anchor of the graph. Each subgraph could then be scheduled independently. We present instead a more efficient algorithm called *iterative incremental scheduling* that solves the relative scheduling problem without decomposing the constraint graph. The algorithm is an extension of the technique used by Liao and Wong [LW83] for layout compaction to support vector solutions. It is guaranteed to find the minimum relative schedule, or detect the presence of inconsistent timing constraints, in polynomial time.

6.3.1 Finding anchor sets

We describe first an algorithm called *findAnchorSet* that finds the anchor sets of the vertices. Each anchor of the graph is propagated to its successors, terminating at the sink vertex. The anchor set $A(v)$ for each vertex $v \in V$ is initialized to null. A vertex v has a counter $ftrav(v)$, initialized to 0, that is used to coordinate the traversal through the graph so that each forward edge in $G_f(V, E_f)$ is traversed exactly once. Note that $G_f(V, E_f)$ is assumed to be acyclic, as described in Chapter 4.

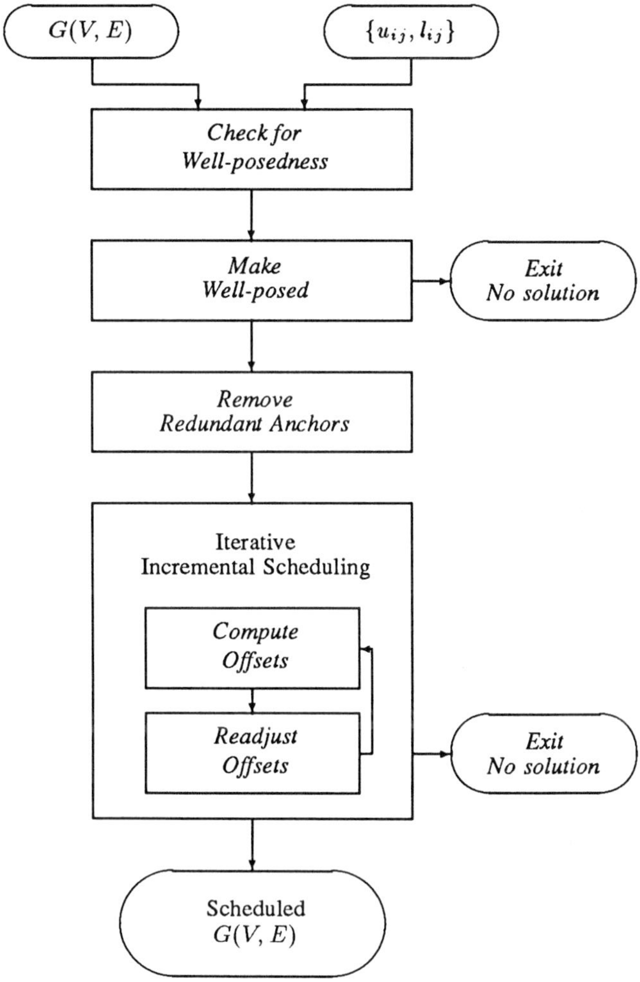

Relative Scheduling Block diagram

Figure 6.10: Block diagram of the relative scheduling approach.

```
findAnchorSet( v, tagSet ) {
        /* increment counter */
        if ( v is not source vertex )
                ftrav(v) ++;
        /* merge tagSet */
        A(v) = A(v) ∪ tagSet;

        if ( ftrav(v) == |pred(v)| ) {
                if ( v is not anchor )
                        for each ( s ∈ succ(v) )
                                findAnchorSet( s, A(v) );
                else
                        for each ( s ∈ succ(v) )
                                if ( w_vs = δ(v) )
                                        findAnchorSet( s, {v} ∪ A(v) );
                                else
                                        findAnchorSet( s, A(v) );
        }
}
```

The function $succ(v)$ returns the set of successors of v. Procedure *findAnchorSet* is applied to the source vertex v_0, where `tagSet` is initialized to $\emptyset$. The worst case complexity of the algorithm is $O(|E_f| \cdot |A|)$ since each forward edge is traversed once, and each traversal requires worst case merging of $|A|$ tags.

6.3.2 Checking well-posed

From Theorem 6.2.2, a constraint graph is well-posed if and only if $A(v_i) \subseteq A(v_j)$ for all $e_{ij} \in E$, *and* the forward constraint graph $G(V, E_f)$ is acyclic. We describe an algorithm called *checkWellposed* that determines whether a constraint graph $G(V, E)$ is well-posed. First, the constraint graph $G_0(V, E)$, where all data-dependent delays are set to zero, is checked for positive cycles to ensure the constraint graph is feasible. The algorithm then checks the anchor sets associated with the ends of every edge in G for containment.

```
checkWellposed( G(V, E) ) {
        /* check for cycles */
        if ( G0(V, E) has positive cycles )
                return unfeasible constraints;
        /* check for containment */
        for each ( e_ij ∈ E_b ) {
                if ( not A(v_i) ⊆ A(v_j) )
                        return ill-posed constraints;
        }
        return well-posed;
}
```

The worst-case complexity of *checkWellposed* is dominated by the check for cycles, which is $O(|V| \cdot |E|)$ [BN86]. The check for containment requires worst-case complexity of $O(|E_b| \cdot |A|)$.

6.3.3 Making well-posed

In some cases, an ill-posed constraint graph $G(V, E)$ can be made well-posed by adding sequencing dependencies to G. Consider for example Figure 6.4(b). The ill-posed constraint can be made well-posed if one adds a sequencing dependency from a_2 to v_i. Although this forces v_i to be serialized with respect to a_2, it is necessary to make the constraint well-posed, i.e. if we are looking for a solution valid under all input sequences. Note that it is not always possible to make an ill-posed constraint well-posed. In particular, if the added sequencing dependency induces a cycle in the forward constraint graph G_f, as in Figure 6.4(a), then the constraint cannot be transformed into a well-posed constraint.

We present the following algorithm, called *makeWellposed*, that guarantees *minimal serialization* in making a constraint graph well-posed.

```
makeWellposed( G(V, E) ) {
      for each ( e_ij ∈ E_b ) do {
            D = {a|a ∈ A(v_i) and a ∉ A(v_j)};
            for each ( a ∈ D )
                  addEdge( a, v_j );
      }
}

addEdge( a, v ) {
      if ( a ∉ A(v) ) {
            if ( v is predecessor of a )
                  stop with ill-posed constraints;
            Add forward edge (a, v);
            Set weight on (a, v) = δ(a);
            A(v) = A(v) ∪ {a};
            for each ( backward edge (v, b) ∈ E_b )
                  addEdge( a, b );
      }
}
```

For every backward edge $e_{ij} \in E_b$, the algorithm first checks if there is an anchor a such that $a \in A(v_i)$ but $a \notin A(v_j)$, i.e. the set D as defined in *makeWellposed.* If no such a exists, then the constraint is well-posed. Otherwise, it attempts to make the constraint well-posed by adding a forward edge from a to v_j. Procedure *addEdge* adds a forward edge from anchor a to all vertices reachable by a path of backward edges from v.

The worst-case complexity of the *makeWellposed* algorithm is $O(|A| \cdot |E_b|^2)$, where $|A|$ is the number of anchors in G. This is because the maximum cardinality of the set D is $|A| - 1$, and the complexity of *addEdge* is $O(|E_b|)$. We prove later that *makeWellposed* yields a minimally serialized well-posed constraint graph, if one exists.

6.3.4 Removing redundant anchors

The equivalence between irredundant start times and start times computed with the full anchor set, as stated by Theorem 6.2.4, makes possible the computation of start times based on irredundant anchors sets. This has advantages of improving the efficiency of the scheduling algorithm and reducing the cost of control.

To compute the irredundant anchor sets, we first identify the relevant anchor sets using an algorithm called *relevantAnchor*, then identify the redundant anchors using algorithm *minimumAnchor* applied to every vertex of the given constraint graph $G(V, E)$.

Find Relevant Anchors The algorithm *relevantAnchor* finds the relevant anchor sets for all vertices. The idea is to propagate an anchor as far as possible on every outgoing path, until either an data-dependent weight edge is encountered, or the sink vertex is reached. Each vertex v_i has a flag $traversed(v_i)$, initialized to **false**, that is used to detect whether the vertex has been previously traversed. The algorithm is applied to every anchor a of the graph with the argument `anchor` set to a.

```
relevantAnchor( v_i, anchor ) {
    /* mark traversed */
    if ( traversed(v_i) )
        return;
    traversed(v_i) = true;
    if ( v_i = anchor ) {
        /* start propagating anchor outwards */
        for each ( outgoing edge e_ij with δ(v_i) )
            relevantAnchor( v_j, anchor );
    } else {
        /* propagate anchor on bounded weight edges */
        R(v_i) = R(v_i)∪ anchor;
        for each ( outgoing bound edge e_ij )
            relevantAnchor( v_j, anchor );
    }
}
```

The worst case complexity of the algorithm is $O(|A| \cdot |V|)$, since each vertex is traversed at most once for each anchor in the graph.

Find Irredundant Anchors The algorithm *minimumAnchor* detects the redundant relevant anchors, given that the relevant anchors have been identified. The algorithm is applied to every vertex v of the graph, where the function $length(a, b)$ returns the longest path from vertices a and b, including backward edges.

```
minimumAnchor( v ) {
    /* for all relevant anchors */
    for each ( relevant anchor r ∈ R(v) ) {
        /* all relevant anchors predecessor to r */
        X = {x|x ∈ R(v), x ∈ A(r)}
        foreach ( x ∈ X )
            if ( length(x, v) ≤ length(x, r) + length(r, v) )
                mark x redundant in R(v);
    }
}
```

The set of unmarked relevant anchors for v form the irredundant anchor set for v, which by Theorem 6.2.6 is the *minimum* anchor set for v. The worst-case complexity of the algorithm is dominated by computing the longest paths, which is $O(|V| \cdot |E|)$. The checking requires $O(|R|^2)$ once the longest path lengths are known, where $|R|$ is the size of the largest relevant anchor set in G.

6.3.5 Iterative incremental scheduling

The scheduling algorithm is performed by iteratively applying two tasks. The first is *incrementally computing the offsets*. The offsets are initially set to zero, and increased incrementally until all the minimum timing constraints implied by the forward edges are satisfied. This is followed by *readjusting offsets* to meet the maximum timing constraints implied by the backward edges. The scheduling algorithm is described below.

```
IncrementalScheduling( G(V, E) ) {
    for (c = 1 to |E_b| + 1) do {
        IncrementalOffset(G_f, v_0);
        E_violate = {e_ij ∈ E_b|violate constraint};
        if ( E_violate = Ø)
            return minimum relative schedule;
        ReadjustOffsets(G(V, E));
    }
    return no schedule;
}
```

The algorithm is an extension of Liao and Wong's algorithm for vectored solutions. Similar techniques have also been proposed by Burns [BN86] and Borriello [Bor88b]. We prove later that the algorithm finds the minimum relative schedule, or detects inconsistent timing constraints by executing at most $(|E_b| + 1)$ iterations. The details of each task are described next.

Incremental offset computation

The offsets are computed by successive approximations. Recall that a forward constraint graph $G_f(V, E_f)$ is acyclic. Therefore, the set of offsets satisfying the minimum timing constraints can be found using the longest path calculation from the anchors to their successors. The edge weights in the constraint graph corresponding to the execution delays of data-dependent delay vertices are set to 0, since the graph is assumed to be well-posed. Note that the iterative incremental scheduling algorithm can also be applied initially to check for feasible constraints.

Let $\Omega^r = \{\sigma_a^r(v_i) | a \in A(v_i), \forall v_i \in V\}$ be the values of offsets after the rth iteration of the scheduling algorithm. Note that in describing the algorithms, the full anchor sets $A(v_i), \forall v_i \in V$ are used. However, we can equally use the irredundant anchor sets without affecting the correctness of the scheduling algorithm. The algorithm *IncrementalOffset* incrementally finds the longest path from the anchors to their successors in the forward constraint graph. Specifically, it takes as input $G_f(V, E_f)$ and the values of the offsets $\sigma_a^r(v_i)$, and computes the new offsets $\sigma_a^{r+1}(v_i)$ as follows:

$$\sigma_a^{r+1}(v_i) = \max_{0 \le j \le n} \{\sigma_a^r(v_j) + \text{ length longest path from } v_j\text{-to-}v_i \text{ in } G_f\}$$

for all vertices v_j that have a path from v_j to v_i in G_f, i.e. $\forall v_j \in pred(v_i)$. Initially, for $r = 0$, the offsets are set to 0. The first invocation of *IncrementalOffset* sets each offset $\sigma_a^1(v_i)$ equal to the length of the longest path from a-to-v_i in G_f. However in a subsequent invocation r, the offset $\sigma_a^r(v_i)$ may be longer than the length of the longest path from a-to-v_i in G_f because of the readjustment strategy to satisfy the maximum timing constraints, described in the next section. A formal description of *IncrementalOffset* is given below.

```
IncrementalOffset( G_f, v ) {
    /* increment counter */
    if ( v is not source vertex )
        ftrav(v) = ftrav(v) + 1;
    if ( ftrav(v) == |pred(v)| ) {
        for ( p ∈ pred(v) )
            foreach ( a ∈ A(p) )
                σ_a^{r+1}(v) = max( σ_a^r(v), σ_a^r(p) + w_pv );
        for ( s ∈ succ(v) )
            IncrementalOffset( G_f, s );
    }
}
```

Procedure *IncrementalOffset* is applied to the source vertex, where $ftrav(v)$ is initialized to 0. Since each edge in E_f is traversed once, and each invocation of the *IncrementalOffset* procedure has worst case complexity of $O(|A|)$. The worst case complexity for finding all longest paths is $O(|A| \cdot |E_f|)$.

Readjusting offsets

After invoking *IncrementalOffset* at the rth iteration, the resulting values of the offsets $\Omega^r = \{\sigma_a^r(v_i) | a \in A(v_i), \forall v_i \in V\}$ satisfy all the minimum constraints implied by the forward edges in G_f. If all the inequalities implied by the backward edges (maximum timing constraints) are satisfied, then $\Omega^r(v_i)$ is the minimum relative schedule and the algorithm terminates. Otherwise, the algorithm successively accesses each backward edge in E_b to test if the maximum timing constraint implied by the edge is violated. There is a constraint violation on a backward edge $e_{ij} \in E_b$ with weight $w_{ij} \leq 0$ if, there exists an anchor a common to both anchor sets $a \in A(v_i) \cap A(v_j)$ such that $\sigma_a^r(v_j) < \sigma_a^r(v_i) + w_{ij}$. If the constraint is violated, then the offset $\sigma_a^r(v_j)$ is increased by the minimum amount to satisfy the inequality constraint:

$$\widetilde{\sigma_a^r}(v_j) \Leftarrow \sigma_a^r(v_i) + w_{ij}$$

The modified offsets $\widetilde{\sigma_a^r}(v_j)$ are then updated in the schedule Ω^r. It is important to note that in the case of well-posed timing constraints, $A(v_i) \subseteq A(v_j)$. After the readjustments, *IncrementalOffset* is reapplied, and the process repeats until all maximum timing constraints due to the backward edges are satisfied. A formal description of *ReadjustOffset* is given below.

```
ReadjustOffset( G(V, E) ) {
        for each ( e_ij ∈ E_b )
                for each ( a ∈ A(v_i) ∩ A(v_j) )
                        if ( σ_a^r(v_j) < σ_a^r(v_i) + w_ij )
                                σ̃_a^r(v_j) = σ_a^r(v_i) + w_ij;
}
```

Complexity of algorithm

We comment now on the total computational complexity of the algorithm. The complexity of the *IncrementalOffset* is $O(|A|\cdot|E_f|)$. The complexity of the readjustment is $O(|A|\cdot|E_b|)$. Therefore, each iteration has computation complexity $O(|A|\cdot|E|)$, proportional to the number of edges in the graph. The *iterative incremental scheduling* algorithm has worst case complexity $O((|E_b|+1)\cdot|A|\cdot|E|)$. Note that in practice the number of backward edges $|E_b|$ and the number of anchors $|A|$ are usually small.

Example of algorithm

We illustrate the application of the algorithm on the graph of Figure 6.11. There are two anchors v_0 and a, the dashed-arcs represent backward edges. The offsets with respect to these two anchors for each vertex are given, where a dash $-$ in the table implies that the corresponding anchor is not in the anchor set of the vertex. For example, a is not in the anchor set of v_6. In the first iteration, the offsets are initialized to zero and computed using longest path search considering only the forward edges. At this point, three backward edges (maximum timing constraints) are violated. We then delay the offsets by the minimum amount to meet the maximum constraints. Consider vertex v_2. Before the readjustment, the values of the offsets are $(2, 1)$. However, it violates the backward edgefrom v_3 to v_2 with weight -1. Therefore, the offsets are adjusted by delaying v_2 by the minimum amount (-1) meet the constraint, i.e. adjust to $(5-1, 4-1) = (4, 3)$. With these offset values, the incremental offset computation is reapplied. In this case, only one backward edge remains violated. The offsets are readjusted again, and offsets recomputed incrementally. The scheduling algorithm terminates with the minimum schedule in the third iteration. The first and second iterations are illustrated in Figures 6.12 and 6.13; the final scheduled solution is shown in Figure 6.14.

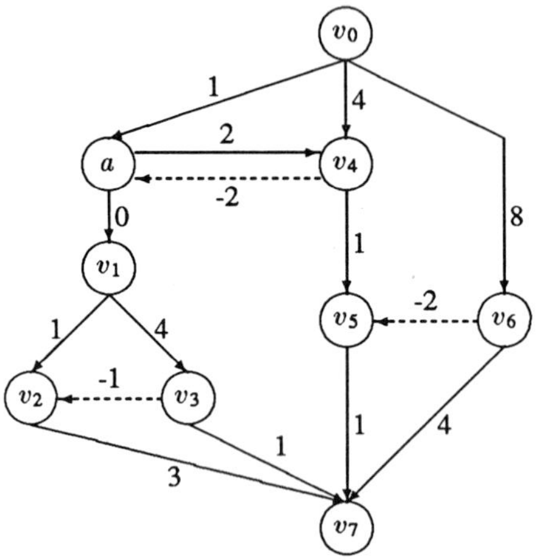

Vertex	*Iteration 1*		*Iteration 2*		*Final*
	Compute σ_{v_0}, σ_a	Readjust σ_{v_0}, σ_a	Compute σ_{v_0}, σ_a	Readjust σ_{v_0}, σ_a	Compute σ_{v_0}, σ_a
v_0	-,-		-,-		-,-
a	1,-	2,-	2,-		2,-
v_1	1,0		2,0		2,0
v_2	2,1	4,3	4,3	5,3	5,3
v_3	5,4		6,4		6,4
v_4	4,2		4,2		4,2
v_5	5,3	6,3	6,3		6,3
v_6	8,-		8,-		8, -
v_7	12,5		12,6		12,6

Figure 6.11: Execution trace for the iterative incremental scheduling algorithm.

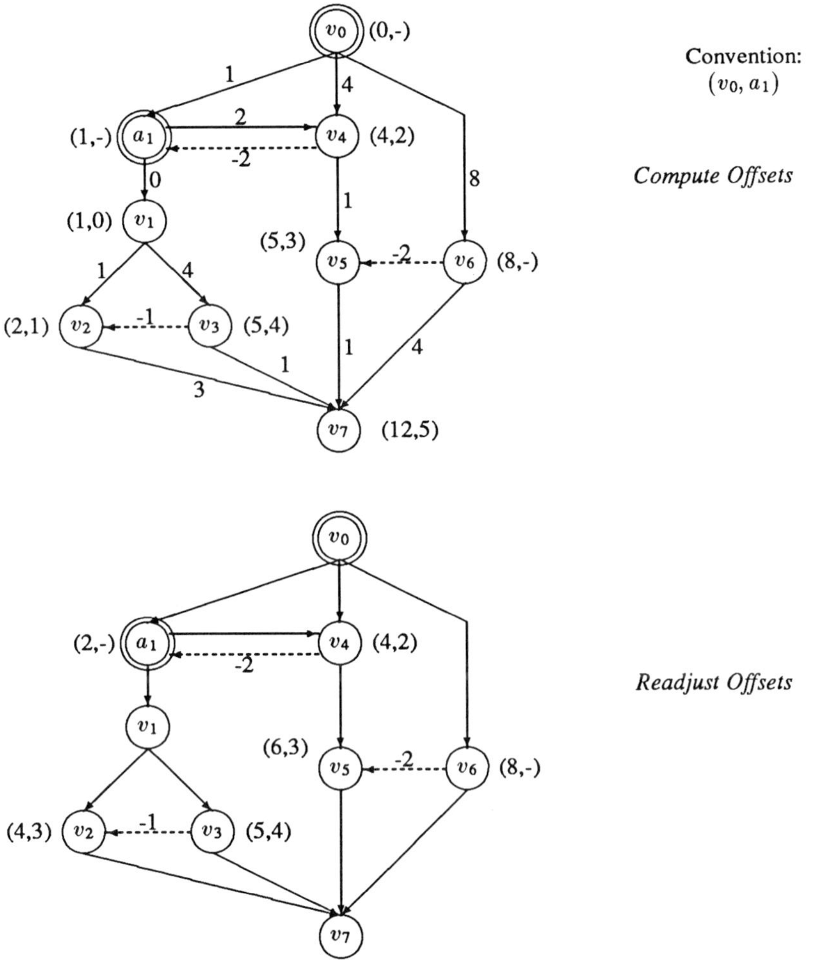

Figure 6.12: First iteration of iterative incremental scheduling algorithm.

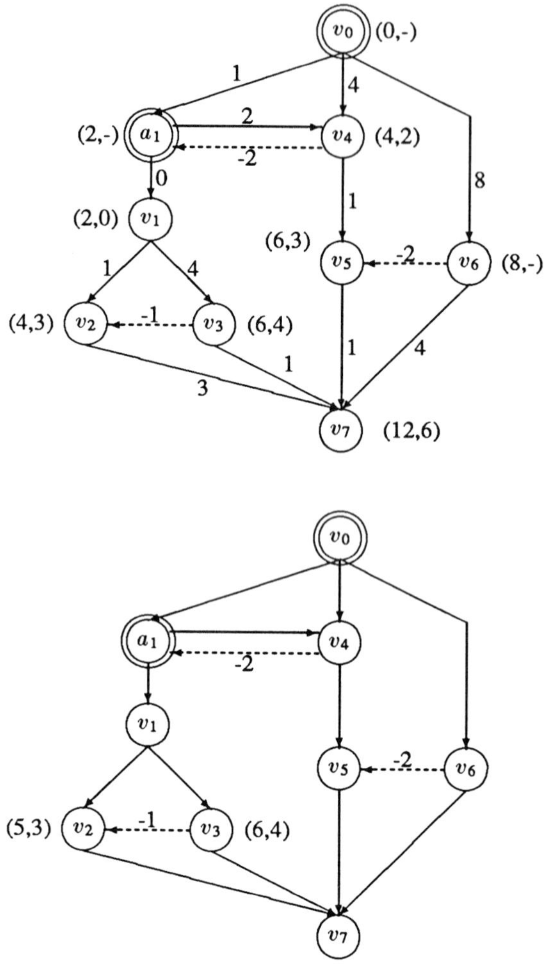

Compute Offsets

Readjust Offsets

Figure 6.13: Second iteration of iterative incremental scheduling algorithm.

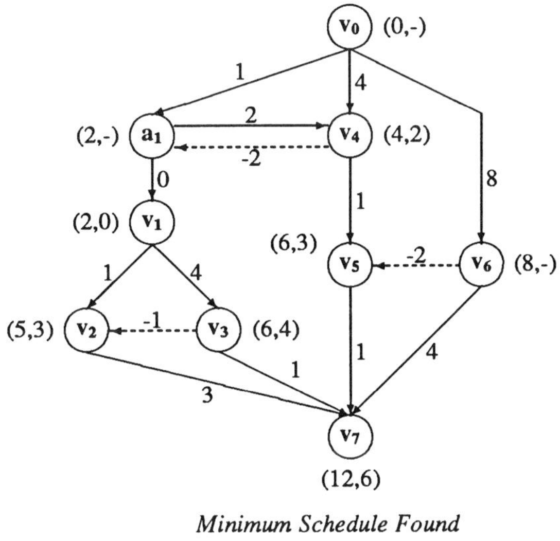

Figure 6.14: Final minimum relative schedule for the example.

6.4 Analysis of Algorithms

We analyze in this section properties of the algorithms presented in Section 6.3. We prove first the *makeWellposed* algorithm can *minimally* serialize an ill-posed constraint graph in attempt to make it well-posed, if a well-posed solution exists. We then prove the *iterative incremental scheduling* algorithm can construct a *minimum relative schedule*, if one exists, in polynomial time.

6.4.1 Analyzing making well-posed

Given an ill-posed constraint graph, we prove in this section that the *makeWell-posed* algorithm yields a minimally serialized constraint graph that is well-posed, if one exists. Let $G(V, E)$ be a constraint graph, where the edges are divided into forward and backward edges $E = E_f \cup E_b$. Then a **serial-compatible graph** of $G(V, E)$, denoted by $G'(V, E'_f \cup E_b)$ is a constraint graph with identical vertex set V and backward edge set E_b, where $E_f \subseteq E'_f$. In other words, a serial-compatible graph G' of G is the original graph G with additional forward edges.

We state first the existence condition for a well-posed solution using the algorithm.

Lemma 6.4.1 *Given a feasible constraint graph G, the makeWellposed algorithm yields a serial-compatible graph of G that is well-posed, if one exists.*

Proof: We show first that if it is possible to make the constraint graph well-posed, *makeWellPosed* can find a solution. By Theorem 6.2.2, it is sufficient to show that *makeWellposed* guarantees the anchor containment criterion, e.g. $A(v_i) \subseteq A(v_j)$ for all $e_{ij} \in E$. We prove first that *makeWellposed* constructs a graph that meets the containment criterion, then we show the graph to be a well-posed serial-compatible graph of the original graph.

The containment criterion is satisfied by definition for forward edges E_f. We consider now backward edges E_b. To make the graph well-posed, for a backward edge e_{ij} that does not satisfy the condition $A(v_i) \subseteq A(v_j)$, it is necessary to add a forward edge e_{aj} to v_j from every anchor $a \in A(v_i)$ but $a \notin A(v_j)$. The weight of the edge w_{aj} is set to $\delta(a)$. Consider now a path of backward edges $(v_i, v_1), (v_1, v_2), \cdots, (v_k, v_j)$. If each backward edge satisfies the anchor containment criterion, i.e. $A(v_i) \subseteq A(v_1), A(v_1) \subseteq A(v_2), \cdots, A(v_k) \subseteq A(v_j)$, then the vertices v_i and v_j also satisfy the anchor containment criterion, i.e. $A(v_i) \subseteq A(v_j)$.

Extending the previous argument, in order for a path of backward edges from v_i to v_j to meet the anchor containment condition, it is necessary to add a forward edge e_{aj} to v_j from every anchor $a \in A(v_i)$ but $a \notin A(v_j)$. Since the anchors from the tail of every backward edge violating the condition are propagated by procedure *addEdge*, and since *addEdge* adds a forward edge from the anchor to the anchor set of *every* vertex reachable by a path of backward edges, the resulting graph satisfies the anchor containment condition. Furthermore, since *makeWellposed* adds new edges to the graph, leaving the original edges unchanged, the resulting graph is a serial-compatible graph of the original graph.

We now show that the algorithm can detect if no well-posed solution exists. From Lemma 6.2.3, G can be made well-posed if and only if no data-dependent length cycle exists. Before adding the edge, the algorithm checks if a cycle would be formed in the graph. If a cycle is formed, then the graph cannot be made well-posed, and the algorithm terminates. If no data-dependent length cycle exists, then *makeWellposed* will not introduce any data-dependent length cycle because of the check in Procedure *addEdge* for whether vertex v is a predecessor of a. Since the algorithm checks the anchor set containment condition for every edge, and since no data-dependent cycle is introduced if the original graph has no data-dependent cycles, the procedure can always find a well-posed solution, if one exists. ||

We now prove the minimum serialization property of *makeWellposed*. A **minimum serial-compatible graph** of $G(V, E)$, denoted by $G'_{min}(V, E'_{min})$, is a serial-compatible graph that is well-posed, and such that the longest path length $length(v_i, v_j)$ $\forall v_i, v_j \in V$ is minimum for all well-posed serial-compatible graphs of G. We state the following theorem.

Theorem 6.4.1 *Given a feasible constraint graph G, the makeWellposed algorithm yields a minimum serial-compatible graph of G, if one exists.*

Proof: By the definition of anchor sets, an anchor a is in the anchor set of a vertex v_i if there exists a path of forward edges from a to v_i with data-dependent length. Without loss of generality, we consider a backward edge $e_{ij} \in E_b$ for which there exists an anchor $a \in A(v_i)$ but $a \notin A(v_j)$. Procedure *makeWellposed* adds a forward edge from a to v_j with data-dependent weight $\delta(a)$.

Note that a is now added to $A(v_j)$, where the maximal defining path $\rho^*(a, v_j)$ trivially reduces to a path containing only the newly created forward edge. Since the original edges remain unchanged, and since the created forward edges satisfy the condition $A(v_i) \subseteq A(v_j)\ \forall e_{ij} \in E_b$ with minimum length (i.e. $|\rho^*(a, v_j)| = 0$, meaning that the length of the longest path from a to v_j excluding the delay $\delta(a)$, is zero), if the resulting graph is well-posed, then it is a minimum serial-compatible graph of G. ||

6.4.2 Analyzing iterative incremental scheduling

The *iterative incremental scheduling* algorithm constructs a minimum relative schedule, or detects the presence of inconsistent timing constraints, with at most $|E_b| + 1$ iterations. This is a very desirable property, since the number of maximum timing constraints $|E_b|$ is in general small. The proof follows the outline in [LW83]. Note that in the sequel the full anchor set $A(v_i)$ for a vertex v_i is used in the computation of the start time and offsets. By Theorem 6.2.4 and Theorem 6.2.6, the result is applicable when the relevant anchor set $R(v_i)$ or the irredundant anchor set $IR(v_i)$ are used instead.

We consider the effect of iterative application of the *iterative incremental scheduling* algorithm. For any integer r, $r \geq 0$, let $\Omega^r = \{\sigma_a^r(v_i) | a \in A(v_i), \forall v_i \in V\}$ denote the offsets after the rth call to the IncrementalOffset procedure. Initially, all offsets $\sigma_a^0(v_i)$ are set to 0. We state the following lemma.

Lemma 6.4.2 *For all $v_i \in V$, the offset $\sigma_a^r(v_i)$ w.r.t. an anchor $a \in A(v_i)$ is equal to the length of a path from a-to-v_i in the constraint graph $G(V, E)$. Furthermore, $\sigma_a^r(v_i) \geq \sigma_a^{r-1}(v_i)$ for any $r \geq 1$.*

Proof: We will prove by induction. After the first call to procedure IncrementalOffset, the offsets $\Omega^1 = \{\sigma_a^1(v_i) | a \in A(v_i), \forall v_i \in V\}$ are equal to the longest paths in G_f from the anchors to their successors. In addition, all offsets $\sigma_a^1(v_i)$ are greater than or equal to 0. Since G_f is a subgraph of the constraint graph G, the assertion holds for $r = 1$.

Assume the assertion is true for $r = k$. This means that for every anchor a in Ω^k, the corresponding offset $\sigma_a^k(v_i)$ is the sum of edge weights on a path from a-to-v_i in G after the kth call to *IncrementalOffset*. Before the next call, the algorithm successively examines each backward edge (v_i, v_j) to see if the inequality constraint is satisfied. Let $w_{ij} \leq 0$ be the weight of a backward edge. Consider an anchor common to both anchor sets, $a \in A(v_i) \cap A(v_j)$. A violation arises if,

$$\sigma_a^k(v_j) < \sigma_a^k(v_i) + w_{ij}$$

whereupon $\sigma_a^k(v_j)$ is set to $\sigma_a^k(v_i) + w_{ij}$. This means that $\sigma_a^k(v_j)$ is assigned a value equal to the length of a path from a-to-v_j consisting of a path a-to-v_i

with length $\sigma_a^k(v_i)$, followed by the edge (v_i, v_j) with length w_{ij}. The increased offset becomes the length of a new path from a to v_j. For the anchors $a \in A(v_j)$ and $a \notin A(v_i)$, the offsets remain the same. Note that $\sigma_a^k(v_j)$ may be moved more than once, each time it is moved to the end of a longer path from a-to-v_j. Therefore, either the offsets remain the same after the readjustment, or they are increased to the length of a longer path.

IncrementalOffset accepts these readjusted offsets as input, and finds the offsets $\Omega^{k+1} = \{\sigma_a^{k+1}(v_i) | a \in A(v_i), \forall v_i \in V\}$ such that, for every anchor $a \in A(v_i)$, the offset $\sigma_a^{k+1}(v_i)$ is

$$\sigma_a^{k+1}(v_i) = \max_{0 \le j \le n} \{ \ \sigma_a^k(v_j) + \text{length longest path from } v_j\text{-to-}v_i \text{ in } G_f\}$$

For $i = j$, the above inequality states that $\sigma_a^{k+1}(v_i) \geq \sigma_a^k(v_i)$ for all anchors $a \in A(v_i)$. Furthermore, Ω^{k+1} consists of offsets that are equal to the lengths of paths from the corresponding anchors $a \in A(v_i)$ to v_i in G. The induction is complete. ||

Now we consider the optimality of the scheduling algorithm. For a constraint graph without positive cycles, define $V(a) \subseteq V$ to be the subset of the vertices in the graph whose anchor set contains a. Specifically, $V(v_0) \equiv V$ since v_0 is included in the anchor set of every vertex. Let $S_k^a, k \geq 0$ be a subset of $V(a)$ such that a vertex v is in S_k^a if, among all the longest weighted paths from a-to-v, the one with the smallest number of backward edges has exactly k backward edges. By definition, S_k^a can be the empty set, and $S_i^a \cap S_j^a = \emptyset$ if $i \neq j$. Let $b = |E_b|$, then for $k > b$, S_k^a is the empty set. Furthermore, the set $\{S_0^a, \cdots, S_b^a\}$ is a partition of $V(a)$, where $V(a) = \bigcup_{i=0}^{b} S_i^a$. Define a number L_a as follows,

$$L_a = \min\{u \mid S_i^a = \emptyset \text{ for } i > u\}$$

This means that L_a is the least number such that, for any vertex $v \in V(a)$, any of the longest weighted paths from a-to-v has no more than L_a backward edges. Furthermore, define L as

$$L = \max\{L_a \mid \forall a \in A\}$$

Obviously, $L \leq b$. We state the following theorem.

Theorem 6.4.2 *Let $G(V, E)$ be a well-posed constraint graph. Then the iterative incremental scheduling algorithm yields the minimum relative schedule after at most $L + 1$ iterations.*

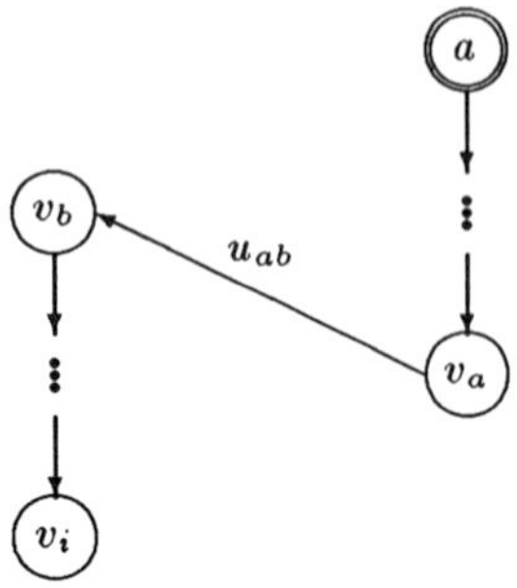

Longest path ρ *from* a-to-v_i

Figure 6.15: The longest path ρ from anchor a to $v_i \in S_k^a$, containing the u_{ab} as the last backward edge on ρ.

Proof: By Lemma 6.4.2, for a vertex $v_i \in V$ in a constraint graph without positive cycles, the offset $\sigma_a(v_i)$ w.r.t. an anchor $a \in A(v_i)$ will remain unchanged by further iterations once it becomes the length of the longest weighted path from a-to-v_i. We will prove by induction that after the rth call to *IncrementalOffset*, $r \geq 1$, the offset $\sigma_a^r(v_i)$ w.r.t. an anchor $a \in A(v_i)$ of a vertex $v_i \in S_{r-1}^a$ becomes the length of its longest weighted path from a-to-v_i in G. Therefore, for all $v_i \in S_{r-1}^a$, the offsets $\sigma_a(v_i)$ are equal to the corresponding minimum offsets. If this assertion is true, then Theorem 6.2.3 implies that the algorithm will terminate and return a minimum relative schedule, taking at most $L + 1$ iterations.

Consider any anchor a of the constraint graph. For the vertices $v \in S_0^a$, the longest path from a-to-v is in G_f because there are no backward edges on the longest paths to these vertices. The first call to *IncrementalOffset* sets $\sigma_a(v)$ to the length of the longest weighted path from a-to-v. Therefore, the assertion is true for $r = 1$.

Now assume the assertion is true for $r = k$. From Lemma 6.4.2, after the kth call to *IncrementalOffset*, a vertex $v \in \bigcup_{p=0}^{k-1} S_p^a$ is the tail of the longest path from anchor a-to-v. Now, let v_i be a vertex $v_i \in S_k^a$ such that the longest path ρ from a-to-v_i contains k backward edges. Let (v_a, v_b) be the last, i.e. k^{th} backward edge on the path ρ with edge weight u_{ab}, as shown in Figure 6.15. The v_b-to-v_i portion of ρ does not have any backward edges by definition. For all longest weighted paths from a to a vertex on the path ρ, the one going through ρ has the least number of backward edges. Therefore, $v_a \in S^a_{k-1}$. By the induction hypothesis, $\sigma^k_a(v_a)$ becomes the length of the longest weighted path from a-to-v_a. When the $(k+1)^{th}$ call to *IncrementalOffset* begins, $\sigma^k_a(v_b)$ has already been set to $\sigma^k_a(v_a) + u_{ab}$, which is also the length of the longest weighted path from a-to-v_b.

After the $(k+1)^{th}$ call to *IncrementalOffset*, we have the following,

$$\sigma^{k+1}_a(v_i) \geq \sigma^{k+1}_a(v_b) + \text{length of the longest path } v_b\text{-to-}v_i$$

The right hand side expression is equal to the length of the longest path a-to-v_i in G. From Lemma 6.4.2, once the offset is equal its longest path length, it will not be increased further. Therefore, all offsets of $v_a \in S^a_{k-1}$ will remain unchanged in later iterations, and $\sigma^{k+1}_a(v_i)$ equals the length of the longest path from a-to-v_i.

For an anchor a, at most $L_a + 1$ iterations are needed to find the minimum relative schedule because $S^a_k = \emptyset,\ k > L_a$. For all anchors, the algorithm will give the minimum relative schedule with at most $L + 1$ iterations. ||

Corollary 6.4.1 *If the constraints implied by the constraint graph* G *are inconsistent, then the algorithm will detect the inconsistency and terminate after* $|E_b|$ *+1 iterations.*

Proof: Assume the constraints are inconsistent, implying a positive cycle exists in the graph. Consider the offset $\sigma_a(v_i)$ w.r.t. an anchor $a \in A(v_i)$ for a vertex v_i on the positive cycle, denoted by $\sigma_a(v_i)$. As *IncrementalOffset* incrementally tries to increase the offsets in order to meet the constraints implied by the forward edges, the readjustment strategy will always increase the value of $\sigma_a(v_i)$. At least one inequality implied by the backward edges will not be satisfied at each iteration. Thus, the algorithm will continue until the $(|E_b|+1)^{th}$ iteration, whereupon the algorithm terminates and returns no schedule. ||

6.5 Summary of Relative Scheduling

Relative scheduling extends the traditional scheduling formulation to support both operations with fixed and data-dependent execution delays. It provides a theoretical framework for analyzing consistency of timing constraints in the presence of these data-dependent delays, using the constraint graph model that is derived from the sequencing graph under detailed timing constraints as the basis for the formulation.

In relative scheduling, the start time of an operation is defined as time offsets with respect to the completion of anchors. Constraints are feasible or well-posed depending on whether they can be satisfied under restricted or general input conditions, respectively. Redundancy of anchors was introduced to simplify the start time of operations by removing redundant anchor dependencies. This can lead to a more efficient control implementation because operations need to be synchronized to a fewer number of signals. Analysis of these properties was presented in this chapter.

This chapter also described efficient algorithms to solve various problems in relative scheduling. Specifically, new algorithms were developed for the following: checking for well-posedness of timing constraints, making constraints well-posed with minimum serialization, removing all redundant anchor dependencies, and finding the minimum relative schedule. All algorithms have polynomial-time complexity and hence can be effectively integrated into the synthesis framework.

The primary objective of relative scheduling is to find a schedule that satisfies the required timing constraints, taking into account data-dependent delays. Within this framework, there are two secondary goals. The first goal is to minimize the hardware latency by finding the minimum schedule under timing constraints, and was the subject of this chapter. The second goal is to minimize the control implementation cost subject to timing constraints; it is the topic of Chapter 9. The relative scheduling framework is also used by the conflict resolution phase in Chapter 7 and the control generation phase in Chapter 8.

Chapter 7

Resource Conflict Resolution

This chapter describes algorithms for resource conflict resolution. Resource conflicts occur when multiple operations activate the same hardware resource simultaneously. When all operations in the hardware model have fixed execution delays, conflict resolution becomes part of the scheduling and resource binding tasks. In particular, operations scheduled to different control steps or belonging to mutually exclusive conditional branches can share their hardware resources. Consider for example the force-directed scheduling technique [PK89b]. Operations with similar resources are first scheduled to reduce their concurrency, then they are bound to hardware resources subject to this schedule. The binding step ensures that no resource conflicts will arise. This approach is, however, restricted to bounded delay operations.

To support data-dependent delay operations in the sequencing graph model, conflict resolution is formulated as the task of *serializing* operations bound to the same resource. This chapter presents a technique called *constrained conflict resolution* that takes as input a sequencing graph with timing constraints and a resource binding. It serializes the sequencing graph to resolve the resource conflicts such that the timing constraints are still satisfied after the serialization. In addition to the support for data-dependent delay operations and detailed timing constraints, this technique uses the topology of the timing constraints to improve the computation time of the resolution algorithm. Resource sharing mutually exclusive conditional branches is also supported in this formulation. Once the graph model has been appropriately serialized, relative scheduling is performed and the corresponding control logic is generated. If the resource conflicts cannot be resolved under timing constraints, then another resource binding is selected

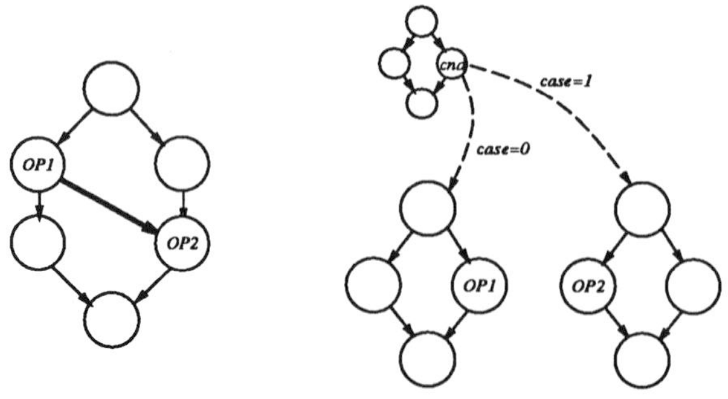

Figure 7.1: Two cases when $op1$ and $op2$ are *implementable:* (a) when they are serialized with each other, or (b) when they reside in mutually exclusive conditional branches.

as candidate for synthesis.

This chapter is presented after relative scheduling because many of the concepts in relative scheduling are used in the conflict resolution formulation. It is organized as follows. Section 7.1 describes the problem formulation that is defined on the constraint graph model. Section 7.2 describes the topology of the timing constraints, introducing the concepts of *orientation* and *polarization* that are used in the resolution algorithms. Section 7.3 presents both exact and heuristic algorithms to resolve conflicts subject to timing constraints. Section 7.4 summarizes the conflict resolution approach in this dissertation.

7.1 Conflict Resolution Formulation

A resource binding is valid if it is possible to resolve its resource conflicts and still satisfy the required timing constraints. For a given resource binding β, recall that an *instance operation set* $O_{(t,i)}$ of β is a subset of vertices that are bound to an allocated resource instance (t, i). Obviously, resource conflicts will occur if the vertices in $O_{(t,i)}$ can execute in parallel. An *implementable* binding is defined as follows.

Definition 7.1.1 *An instance operation set $O_{(t,i)}$ is* **implementable** *if the elements of $O_{(t,i)}$ are disjoint in time, i.e. they do not execute concurrently. Given a*

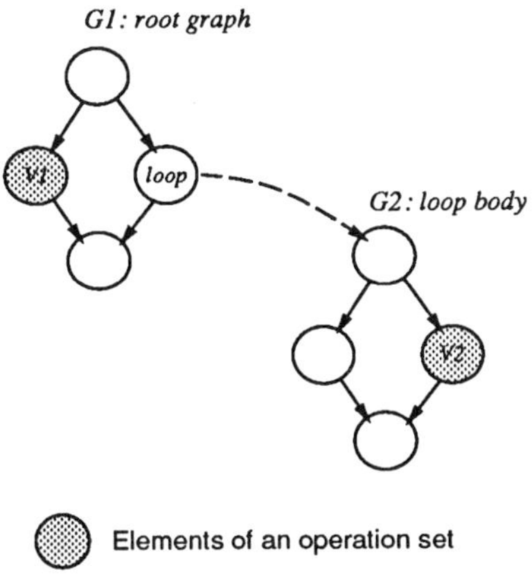

Figure 7.2: Illustrating hierarchical nature of instance operation sets: G_1 is the main graph containing v_1, and G_2 is the loop body containing v_2.

binding β of a constraint graph $G(V, E)$, G is **implementable** *if every instance operation set in β is implementable.*

From Section 5.2, the concurrency factor of an implementable instance operation set is 1 and the conflict degree of an implementable binding is 0.

Two operations $op1$ and $op2$ are disjoint in time if one of the two following conditions holds: (1) $op1$ is serialized with respect to $op2$ in the graph, such that $op1$ can execute only if $op2$ has completed execution or vice versa, and (2) $op1$ and $op2$ each belong to different mutually exclusive branches of a conditional. The two cases are illustrated in Figure 7.1. Since the conditional branching structure cannot be arbitrarily altered without changing the external behavior of the model, we resolve resource conflicts by serializing operations.

The example in Figure 7.1 illustrates the hierarchical control-flow of the sequencing graph model. In particular, elements of an instance operation set $O_{(t,i)}$ may not all belong to the same sequencing graph. Consider the example in Figure 7.2 of a control-flow hierarchy consisting of two graphs G_1 and G_2; G_1 is the root graph and G_2 is the body of the loop in G_1. Shaded vertices represent elements of an instance operation set. Note that elements of an instance operation set may belong to different sequencing graphs.

To address this issue, conflict resolution for an instance operation set $O_{(t,i)}$ in a sequencing graph G_M is performed hierarchically in a bottom-up manner. A *candidate operation set* $O_{(t,i)}(G)$ for each graph G in the cf-hierarchy G^*_M is identified as candidates to be serialized. A vertex v is a candidate if v belongs to the instance operation set $O_{(t,i)}$ or if one or more elements of $O_{(t,i)}$ belong to graphs in the cf-hierarchy induced by v. Returning to Figure 7.2, conflict resolution is first applied on G_2, where there is a single candidate v_2. Moving upwards to the main graph G_1, there are two candidates corresponding to v_1 and the loop vertex.

We consider in the rest of this chapter a single constraint graph model G that is derived from a sequencing graph with timing constraints, where conflict resolution has been performed on all graphs in its cf-hierarchy G^*. Therefore, the term "instance operation set" in the sequel refers to the candidate operation set of $O_{(t,i)}$ with respect to G.

7.1.1 Objective in conflict resolution

The objective in conflict resolution is to resolve the conflicts among elements of a candidate operation set $O_{(t,i)}(G)$, which is derived from a resource binding β and a constraint graph model $G(V, E)$. An *ordering* of the instance operation set is defined as follows.

Definition 7.1.2 *An* **ordering** *of an instance operation set* $O_{(t,i)}(G)$, *denoted by* $\langle o_1, o_2, \ldots, o_k \rangle$ *where* $k = |O_{(t,i)}(G)|$, *is a serialization of the vertices of* $O_{(t,i)}(G)$ *in* $G(V, E)$ *such that in the resulting constraint graph,* o_j *is a predecessor to* o_{j+1}, $1 \leq j \leq k - 1$.

The activation of the vertex $o_j \in O_{(t,i)}$ in an ordering must depend on the *completion* of the preceding vertex $o_{j-1} \in O_{(t,i)}$ in the ordering. An ordering for an instance operation set $O_{(t,i)}(G)$ is a sufficient condition to ensure that $O_{(t,i)}(G)$ is implementable. It is a *valid ordering* if the resulting serialized graph G satisfies the timing constraints, i.e. it is well-posed according to Chapter 6.

7.2 Constraint Topology

This section analyzes the topology of timing constraints in a constraint graph $G(V, E)$. We describe several concepts that are used in the conflict resolution formulation. Let the target instance operation set $O_{(t,i)}(G)$ be denoted by $O \subseteq V$, where we dropped the terms (t, i) and G for conciseness. The instance operation set O consists of $k = |O|$ vertices, denoted by $o_i, i = 1, \ldots, k$. Each vertex $o_i \in O$ has an associated execution delay $\delta(o_i)$ that can be fixed or data-dependent. In the simplistic case of flat graphs, all elements of O are call vertices to the same model; therefore, they have identical execution delays. However, for hierarchical graphs, unequal execution delays may result. This case is illustrated in Figure 7.2, where the candidate operation set in G_1 consists of v_1 and the loop vertex; they have different execution delays. In fact, the loop vertex has data-dependent execution delay.

A cycle in the constraint graph represents a cyclic timing relationship among a set of vertices. From the previous chapter on relative scheduling, a violation of timing constraints can occur if the constraint graph is *unfeasible* or if the constraint graph is *ill-posed.* Theorem 6.2.1 states that a constraint graph is feasible if and only if no positive cycle exists in G assuming data-dependent delays are set to zero. Lemma 6.2.3 states that a constraint graph can always be made well-posed if and only if no data-dependent length cycles exist in G. In both cases, timing constraint violation occurs in the presence of cycles in the graph model. Based on this observation, we partition the elements of the instance operation set by introducing the concept of *operation clusters*, defined below.

Definition 7.2.1 *An* **operation cluster** $\mathcal{C}$ *of an instance operation set* O *is the maximal subset of vertices in* O *that is strongly connected, i.e. there exists a directed path between every pair of vertices in the operation cluster.* $|\mathcal{C}|$ *denotes the cardinality of* $\mathcal{C}$.

Theorem 7.2.1 *A partial order exists among the operations clusters of an operation set.*

Proof: Elements of an instance operation set are strongly connected in the constraint graph. Since strong connectivity is an equivalence relation, two operation clusters cannot be connected by a cycle. This is the definition of partial order. ||

The set of operation clusters is denoted by $\Pi = \{\mathcal{C}_i, i = 1, \ldots, |\Pi|\}$, where $|\Pi|$ is the number of operation clusters in O. The operation clusters form a partition over the elements of O because the property of strong connectivity is an equivalence relation. We illustrate the concept in Figure 7.3, where the dotted arcs represent backward edges with negative weights and the solid arcs represent forward edges with positive weights. There are two operation clusters $C_1 = \{A, B, C\}$ and $C_2 = \{D, E\}$ in the example. A partial order is formed over the two clusters, i.e. from C_1 to C_2.

This partial order over the operation clusters provides the basis for a conflict resolution strategy based on decomposition. Specifically, the problem of finding a valid ordering for an instance operation set is divided into two steps:

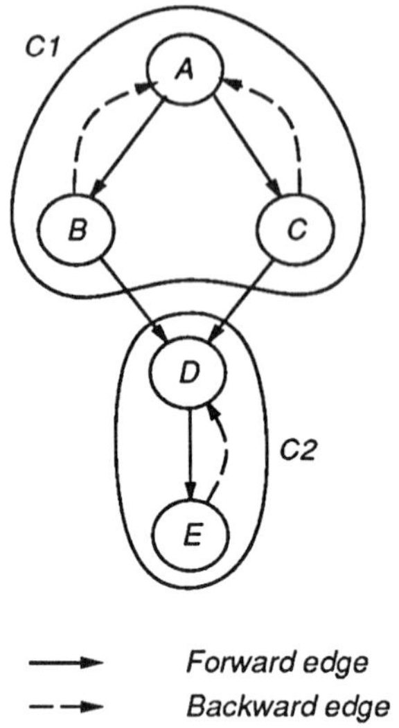

Figure 7.3: Example of an instance operation set with 5 vertices $\{A, B, C, D, E\}$. Two operation clusters are formed: $\mathcal{C}_1 = \{A, B, C\}$ and $\mathcal{C}_2 = \{D, E\}$.

- *Ordering among the operation clusters:* Find a linear order of operation clusters that is compatible with the induced partial order in Π, and
- *Ordering within each operation cluster:* Find a valid ordering for the vertices within each operation cluster.

We state the following theorem.

Theorem 7.2.2 *If valid orderings exist for the vertices inside each operation cluster $\mathcal{C}_i \in \Pi, i = 1, \ldots, |\,\Pi\,|$, then any ordering of operation clusters that is compatible with the partial order induced by Π is a valid ordering for O.*

Proof: Assume each operation cluster has a valid ordering. Since clusters are not connected by a cycle, the serialization of one cluster does not affect any cyclic constraints of the other operation clusters. Since each cluster is ordered and no constraints are violated by ordering among the clusters, the resulting ordering is valid for the entire instance operation set. ||

With Theorem 7.2.2, the problem of finding a valid ordering for an instance operation set O has been reduced to the problem of finding a valid ordering for the elements of an operation cluster $C_i \in \Pi$. The formation of operation clusters is strongly dependent on the extent to which operations are related by timing constraints. By linking the complexity of the resolution effort with the complexity of operation clusters, we take advantage of the *topology* of constraints in finding an efficient serialization.

7.2.1 Orientation and polarization

We introduce in this section the concepts of *orientation* and *polarization* of an operation cluster. For conciseness, we consider one operation cluster C that contains $|C|$ vertices, i.e. $C = \{c_i | i = 1, \ldots, |C|\}$.

We make the following assumptions. First, the cardinality of the operation cluster must be greater than one ($|C| > 1$), since otherwise the ordering is trivial. Second, each vertex $c_i \in C$ must either be a data-dependent delay operation (i.e. anchor) or have non-zero fixed execution delay, i.e. $\delta(c_i) > 0$. Note that registers have already been introduced prior to conflict resolution to latch the outputs of the shared resource. For example, the execution delay for shared calls to a combinational adder is 1 cycle because of the latching delay.

The above assumptions translate into the following two requirements on any cycle connecting the elements of the cluster.

- *Bounded length.* Any cycle must have bounded length, i.e. no edge with data-dependent weight exists on the cycle. This ensures that the resulting serialized graph can be made well-posed (Lemma 6.2.3).

- *Negative length* . Assuming the cycle length is bounded, the cycle must have zero or negative length. This ensures that the resulting graph is feasible (Theorem 6.2.1).

Orientation

Paths among elements in a cluster are categorized according to the sign of their lengths. The reason is because conflict resolution serializes the operations. Therefore, new paths are introduced having positive lengths. We determine the *orientation* of the vertices with respect to positively weighted paths to prune some of the invalid ordering possibilities. Specifically, an orientation is defined as follows.

Definition 7.2.2 *The* **orientation** *of an operation cluster C is a relation $\mathcal{P}_C$: $C \times C$, where $(x, y) \in \mathcal{P}_C$ if the length of the longest path from x to y is greater than or equal to zero, i.e. $length(x, y) \geq 0$.*

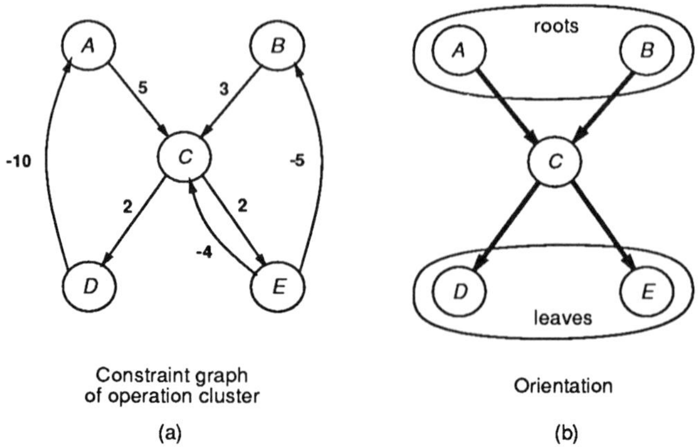

Figure 7.4: Example of (a) an operation cluster with 5 vertices $\{A, B, C, D, E\}$, (b) the corresponding orientation. The roots are $\{A, B\}$, and the leaves are $\{D, E\}$.

Consider for example an operation cluster containing 5 vertices $\{A, B, C, D, E\}$. The orientation is obtained by retaining only the positively weighted edges. This is shown in Figure 7.4.

Intuitively, an orientation of a constraint graph represents a partial order among the vertices that must be satisfied if the resulting serialized graph is to be valid. For example, $(x, y) \in \mathcal{P}_{\mathcal{C}}$ implies that the serialization $y \Rightarrow x$ will violate the timing constraints, i.e. form a positive cycle. The induced partial order in $\mathcal{P}_{\mathcal{C}}$ defines a predecessor-successor relationship among the elements of $\mathcal{C}$. A vertex $c_i \in \mathcal{C}$ is a predecessor to another vertex $c_j \in \mathcal{C}$ if there exists a pair $(c_i, c_j) \in \mathcal{P}_{\mathcal{C}}$; successors are defined in a similar manner.

Based on this predecessor-successor relation, the **leaves (roots)** of an orientation $\mathcal{P}_{\mathcal{C}}$ are the subset of elements of $\mathcal{C}$ that have no successors (predecessors). They are denoted by $\mathcal{P}_{\mathcal{C}}^{leaf}$ and $\mathcal{P}_{\mathcal{C}}^{root}$, respectively. Returning to the example in Figure 7.4, the roots of the orientation are $\{A, B\}$ and the leaves are $\{D, E\}$.

Polarization

It is straightforward to show that any valid ordering of $\mathcal{C}$ must be compatible with the partial order induced by the orientation $\mathcal{P}_{\mathcal{C}}$. This observation implies that the first element of any valid ordering must be a root, and similarly the last element must be a leaf. For a root-leaf pair $(r, l) : r \in \mathcal{P}_{\mathcal{C}}^{root}, l \in \mathcal{P}_{\mathcal{C}}^{leaf}$, we can

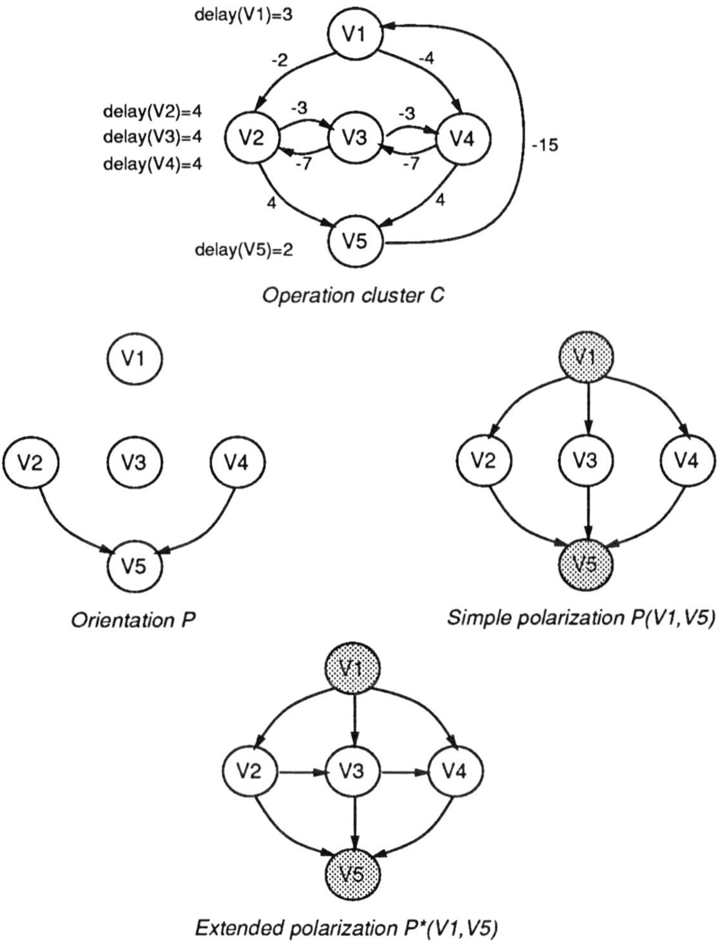

Figure 7.5: Illustrating an operation cluster and its orientation $\mathcal{P}$, a simple polarization $\mathcal{P}(v_1, v_5)$, and the extended polarization $\mathcal{P}^*(v_1, v_5)$.

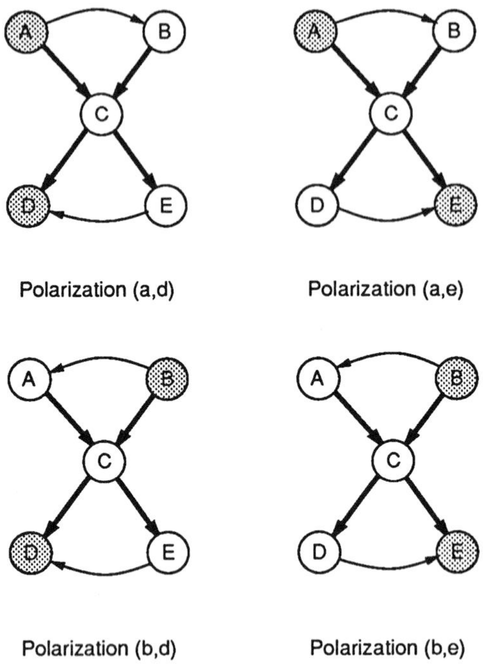

Figure 7.6: Four possible polarizations for an operation cluster where bold arcs represent the orientation.

make the orientation *polar* (single-source and single-sink) by serializing from r to all other vertices and from all vertices to l. We formalize this observation in defining a *polarization.*

Definition 7.2.3 *A* **simple polarization** *with respect to* $r \in \mathcal{P}_C^{root}$ *and* $l \in \mathcal{P}_C^{leaf}$ *of an orientation* $\mathcal{P}_C$, $r \neq l$, *denoted by* $\mathcal{P}_C(r,l)$, *is the relation that is derived from union of* $\mathcal{P}_C$ *with the relations* $(r,v), \forall v \neq r$ *and* $(w,l), \forall w \neq l$. *An* **(extended) polarization,** *denoted by* $\mathcal{P}_C^*(r,l)$, *is* $\mathcal{P}_C(r,l)$ *extended with the all pairs* (v,w) *such that* $length(w,v) + \delta(v) > 0$.

The reason for disallowing the serialization from w to v if the condition $length(w,v) + \delta(v) > 0$ holds is to avoid creating a positive cycle. Figure 7.5 shows an operation cluster of 5 vertices $\{v_1, v_2, v_3, v_4, v_5\}$. Vertices v_2, v_3 and v_4 are connected with one another by negatively weighted edges representing maximum timing constraints, i.e. $w_{v_2,v_3} = -3$ means that v_2 can start no more than

3 cycles after the activation of v_3. The orientation $\mathcal{P}$ is the subgraph induced by the positive weighted edges, where the roots consist of $\mathcal{P}^{root} = \{v_1, v_2, v_3, v_4\}$ and the leaves consist of $\mathcal{P}^{leaf} = \{v_1, v_3, v_5\}$. A simple polarization $\mathcal{P}(v_1, v_5)$ adds edges from v_1 to all remaining vertices and from all non-leaf vertices to v_5. The extended polarization $\mathcal{P}^*(v_1, v_5)$ considers in addition the value of the negatively weighted edges. For example, $(v_2, v_3) \in \mathcal{P}^*(v_1, v_5)$ because a positive cycle is formed if v_3 with execution delay of 4 is serialized to v_2.

Theorem 7.2.3 *If a cycle exists in a polarization $\mathcal{P}_C^*(r,l)$, then no valid ordering exists that is compatible with the polarization.*

Proof: A pair (x, y) in the polarization implies a precedence relationship between x and y, i.e. x must be serialized *before* y. Assume the presence of a cycle in the graph, denoted by $(x, y_1), (y_1, y_2), \ldots, (y_k, x)$. By transitivity of the precedence relationship, the cycle implies that x must be serialized with respect to x, which is inconsistent. Therefore, since any serialization must be compatible with the polarization, no valid ordering exists if a cycle exists in the graph. ||

Any valid ordering of an operation cluster must be compatible with one of its polarizations. There is a finite number of polarizations for a given orientation. The total number of possible polarizations for an orientation $\mathcal{P}_C$ is given by the expression:

$$\#\text{polarization} = |\mathcal{P}_C^{root}| \cdot |\mathcal{P}_C^{leaf}| - |\mathcal{P}_C^{root} \cap \mathcal{P}_C^{leaf}|$$

where the $|\mathcal{P}_C^{root} \cap \mathcal{P}_C^{leaf}|$ term corresponds to the isolated vertices in the orientation.

Figure 7.6 shows an operation cluster with 5 vertices. The bold arcs are due to the orientation and the shaded vertices denote root and leaf vertices in a polarization. There are $2 \cdot 2 - 0 = 4$ polarizations for this cluster. The concept of polarizations allows us to prune the search for a valid ordering. Since the simple polarization $\mathcal{P}_C(r,l)$ is a restriction of the polarization $\mathcal{P}_C^*(r,l)$, we use strictly $\mathcal{P}_C^*(r,l)$ in the rest of the chapter.

7.2.2 Properties of polarizations

This section describes two theorems related to polarizations that are used as filters to speed the search for a valid ordering. The first theorem is related to the presence and position of anchors in a given polarization.

Theorem 7.2.4 *For a polarization $\mathcal{P}^*_{\mathcal{C}}(r,l)$, if any non-leaf vertices in $\mathcal{C}$ is an anchor, then no valid ordering exists for the polarization.*

Proof: Assume there exists a non-leaf vertex $v \neq l$ with data-dependent execution delay. A valid ordering of $\mathcal{P}^*_{\mathcal{C}}(r,l)$ implies that v must be serialized with respect to l. Since v has data-dependent execution delay, the serialization requires introducing an edge with data-dependent weight to the constraint graph. The vertices in $\mathcal{C}$ are however strongly connected, meaning that a data-dependent length cycle has been formed. This means the timing constraints cannot be satisfied, and no valid ordering exists. ||

The following theorem provides an effective and exact pruning measure that is used in the exact conflict resolution algorithm, described in Section 7.3.2. The theorem states that the sum of the execution delays of the vertices to be serialized ($\sum_{v \in \mathcal{C}, v \neq l} \delta(v)$) must not exceed the allowed maximum timing constraint from l to r, i.e. $length(l,r)$.

Theorem 7.2.5 *Consider a polarization $\mathcal{P}^*_{\mathcal{C}}(r,l)$. If the following condition holds:*

$$length(l,r) + \sum_{v \in \mathcal{C}, v \neq l} \delta(v) > 0$$

then no valid ordering exists for the polarization.

Proof: A valid ordering within an operation cluster implies that all vertices are serialized to form a chain. Given a polarization (r,l), r is the first element of the chain and l is the last element of the chain. The minimum length of such a chain is equal to the sum of the execution delays of the vertices excluding the leaf l, i.e. $\sum_{v \in \mathcal{C}, v \neq l} \delta(v)$. A necessary condition for a valid ordering is that no positive cycles are formed in the resulting constraint graph. Consider the cycle formed by the chain and the backward path from l to r, the length of the latter is denoted by $length(l,r)$. If the cycle has positive length, then the resulting graph is invalid and no valid ordering exists. ||

7.3 Algorithms for Conflict Resolution

Algorithms for conflict resolution is presented in this section. The input is a resource binding β consisting of a number of instance operation sets. The instance operation sets in β are selected in turn. For a given instance operation set O, its operation clusters are first identified using standard graph techniques such as cycle detection or path tracing. The following steps are then performed for each operation cluster C_i in O:

1. *Identify the orientation* $\mathcal{P}_C$. The orientation is obtained by categorizing the edges based on the sign of their weights. The roots $\mathcal{P}_C^{root}$ and leaves $\mathcal{P}_C^{leaf}$ of the orientation are identified.

2. *Select a polarization* $\mathcal{P}_C^*(r, l)$. A particular polarization with root r and leaf l is selected. If a cycle exists in the polarization or if the polarization violates the condition in Theorem 7.2.4, then it is discarded and another polarization is selected. If all polarizations are invalid, then the resource conflicts for the given operation cluster cannot be resolved under timing constraints.

3. *Apply heuristic ordering algorithm.* A polynomial-time complexity heuristic algorithm is performed to find a valid ordering with the goal of minimizing the latency of the resulting hardware. If a solution is found, another cluster is selected as candidate and the steps are repeated until all clusters have been ordered. Otherwise, the exact ordering algorithm in the next step is performed.

4. *Apply exact ordering algorithm.* A branch-and-bound ordering algorithm is applied if the heuristic fails to find a solution. The exact algorithm is guaranteed to find a solution if one exists. Theorem 7.2.5 is used as the cost function to prune the branch-and-bound search.

After the operations within each clusters have been serialized, the clusters are linearly ordered compatible with the original partial order. This linear order can be constructed in linear-time with respect to the number of clusters. The order in which the clusters are visited and resolved is important. Therefore, if the above steps fail to find a valid ordering, then this order can be changed and the above steps repeated. By Theorem 6.2.1 and Lemma 6.2.3, a solution to conflict resolution implies that a valid relative schedule exists. In this case, the binding is known to be valid and the iterative incremental scheduling algorithm can be performed to compute the time offsets.

The reason for using both heuristic and exact ordering strategies is due to the complexity of the conflict resolution problem. The problem of finding a valid ordering can be cast as the problem of "sequencing with release times and deadlines", which has been shown to be NP-complete in the strong sense [GJ79]. There have been extensive research on techniques to solve machine shop scheduling and sequencing [Fre82], including linear programming, branch-and-bound, and heuristic programming approaches. Heuristic algorithms are efficient when they can find a solution. Unfortunately, if they fail to find a solution, then it is not possible to guarantee that no solution exists due to the heuristic nature of the formulation.

The above strategy can be used to find a conflict resolution that satisfies the timing constraints, if one exists. For designs with large design space, the exact ordering algorithm can be skipped to expedite the search for a valid conflict resolution; however, in this case it is not guaranteed that a solution can be found even if one exists. Given a polarization, Section 7.3.1 describes the heuristic ordering algorithm and Section 7.3.2 describes the exact ordering algorithm.

7.3.1 Heuristic ordering search

Given a polarization $\mathcal{P}_C^*(r,l)$ of an operation cluster C, a heuristic ordering algorithm is described in this section. The algorithm, called *Heuristic_order*, tries to minimize the latency of the resulting serialized graph. If no valid ordering is found, then another polarization is selected. If the heuristic algorithm fails for every polarization, the exact ordering algorithm in the next section is applied.

This procedure greedily constructs the ordering starting from the leaf element l of the polarization. The variable $current$ keeps a pointer to the most recently ordered element, initialized to l. At each point in constructing the ordering, a set of possible candidates is identified, where the candidates correspond to the operations in the cluster that are compatible with the partial order induced by the polarization. For each candidate v_y, a *slack* value $\int(v_y)$ is defined as follows:

$$\begin{aligned} \int(v_y) \;\; = \;\; -[\; & \max\{\delta(v_y), length(v_y, current)\} \\ & + length(current, l) + length(l, v_y) \;\;] \end{aligned}$$

The computation for $\int(v_y)$ is composed of three path segments: from v_y to $current$ after the serialization, from $current$ to the leaf l, and from the leaf l to v_y. From Theorem 7.2.4, v_y must be a fixed delay vertex. After serializing v_y with respect to $current$, the length of the longest path between them is the maximum of the length $\delta(v_y)$ of the serializing edge and the previous longest path length. Note that by definition of clusters the longest path is defined between every pair of vertices in a cluster.

Intuitively, the slack is a measure of the length of the longest cycle that would be formed if v_y is selected and serialized as the next element in the ordering Ord. It must always be positive since otherwise the serialization is not valid. Among the possible candidates, the one with *minimum* slack is selected

```
Procedure Heuristic_order( polarization P*_C(r,l) ) {
        /* Start with the leaf l */
        Ord = ⟨l⟩;
        current = l; /* construct ordering upwards */
        while ( unordered candidates exist ) {
                Candid = compatible(Ord);
                /* select most constrained candidate */
                v_y = Select arg min_{z∈Candid} {f(z)};
                Add v_y to the ordering Ord and serialize graph;
                Recompute all-pairs longest path;
                /* check timing constraints */
                if (positive cycle formed)
                        return no valid ordering found;
                current = v_y;
        }
        return valid ordering Ord;
}
```

Figure 7.7: Heuristic ordering algorithm *Heuristic_order*.

as the next element in the ordering. The procedure to incrementally construct an ordering $Ord = \langle \ldots \rangle$ is described in Figure 7.7.

The routine $compatible(Ord)$ returns a set of candidates with respect to a partial ordering Ord. To define it, we first augment the original polarization $\mathcal{P}^*_{\mathcal{C}}(r,l)$ with the partial ordering $Ord = \langle Ord_i, \ldots, Ord_{|\mathcal{C}|} \rangle$ by adding the relations $\{(Ord_j, Ord_{j+1}), i \le j \le |\mathcal{C}|-1\}$ to $\mathcal{P}^*_{\mathcal{C}}(r,l)$. An unordered element v_c is in $compatible(Ord)$ if there exists no other candidate $w_c \in compatible(Ord)$ such that the relation (v_c, w_c) is in the augmented polarization defined earlier.

At each iteration of the loop, the graph is serialized with respect to the constructed partial ordering. This ordering is constructed incrementally until the root r is reached. At each iteration, the serialized graph is checked for consistency. Consistency analysis involves computing the longest path lengths between pairs of operations, which using Floyd's algorithm requires complexity $O(|\mathcal{C}|^3)$. Therefore, the overall procedure has $O(|\mathcal{C}|^4)$ complexity.

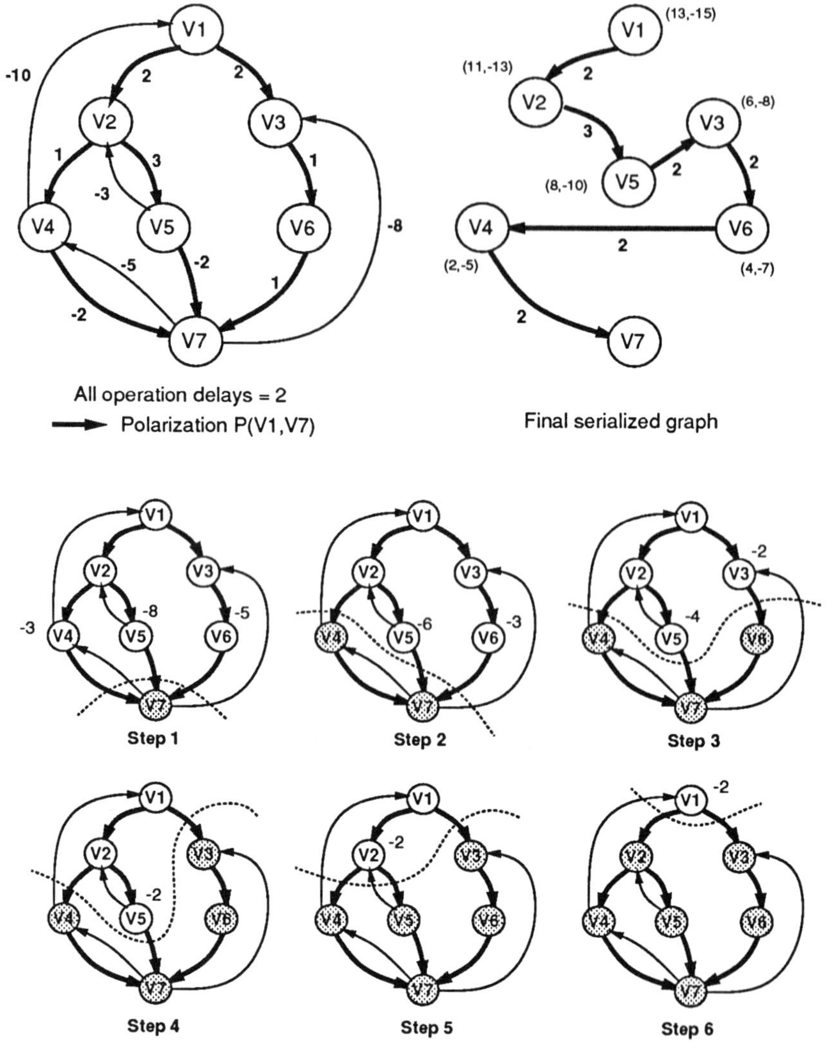

Figure 7.8: Example of the heuristic ordering algorithm applied to a cluster with 7 vertices.

Example. We illustrate the application of procedure *Heuristic_order* in Figure 7.8 to an operation cluster consists of 7 vertices $\{v_1, \ldots, v_7\}$, starting with the polarization $\mathcal{P}^*(v_1, v_7)$. The partial order Ord is constructed from the leaf v_7 upwards to the root v_1. At step 1, the candidates based on the partial order of the polarization (represented by bold arcs) are $\{v_4, v_5, v_6\}$. The slacks for these candidates are:

$$\begin{aligned} \int(v_4) &= -(\max(2,-2)+0+(-5)) &= 3 \\ \int(v_5) &= -(\max(2,-2)+0+(-10)) &= 8 \\ \int(v_6) &= -(\max(2,1)+0+(-7)) &= 5 \end{aligned}$$

Vertex v_4 has minimum slack and hence is added to the partial ordering $Ord = \langle v_4, v_7 \rangle$. The graph is serialized accordingly. At step 2, the candidates are $\{v_5, v_6\}$. The slacks for these candidates are:

$$\begin{aligned} \int(v_5) &= -(\max(2,-2)+2+(-10)) &= 6 \\ \int(v_6) &= -(\max(2,-4)+2+(-7)) &= 3 \end{aligned}$$

Vertex v_6 has minimum slack and is added to $Ord = \langle v_6, v_4, v_7 \rangle$. The algorithm repeats until the root vertex v_1 is reached. The final order $\langle v_1, v_2, v_5, v_3, v_6, v_4, v_7 \rangle$ results in a valid constraint graph.

7.3.2 Exact ordering search

The heuristic ordering strategy in the previous section may fail to find a valid ordering in some cases. This section presents an *exact* ordering algorithm based on branch-and-bound called *Exact_order* is performed to find a valid ordering for a given polarization $\mathcal{P}^*_{\mathcal{C}}(r, l)$. If a valid ordering is not found for this polarization, the algorithm is applied to another polarization. If a valid ordering is not found for any polarization, then it is not possible to resolve the conflicts in the operation cluster $\mathcal{C}$.

This recursive algorithm constructs an ordering incrementally, starting from the leaf l of the polarization. The partial ordering that is being constructed is denoted by $Ord = \langle Ord_i, \ldots Ord_{|\mathcal{C}|} \rangle$; the index i is the index of the current element in the partial ordering, $1 \leq i \leq |\mathcal{C}|$. Note that the first and last elements of Ord are the root and leaf vertices, respectively. The procedure is described in Figure 7.9, where the routine $compatible(Ord)$ is the same as in the previous section.

The ordering is complete when $i = 1$, whereupon the procedure records the valid ordering Ord and returns TRUE. Otherwise, one of the candidates in the

```
Procedure Exact_order( partial ordering Ord, current index i) {
    if ( i = 1 ) {
        Record valid ordering Ord;
        return TRUE;
    }
    /* try each compatible candidates */
    foreach ( z ∈ Compatible(Ord) ) {
        Append z to the ordering Ord;
        /* prune based on cost */
        if ( cost(Ord_i) + length(l, r) < 0 ) {
            Serialize graph subject to Ord;
            if ( resulting graph valid )
                if ( Exact_order( Ord, i - 1) = TRUE )
                    return TRUE;
        }
    }
    /* backtrack */
    return FALSE;
}
```

Figure 7.9: Exact ordering algorithm *Exact_order*.

set returned by $compatible(Ord)$ is added to Ord. For each candidate, pruning is performed to filter out candidates that will violate timing constraints. The pruning strategy is based on defining for the subsequence $\langle Ord_i, \ldots, Ord_{|C|}\rangle$ a *cost function*, denoted by $cost(Ord_i)$ representing a bound on the *length of the longest path* from the root vertex r to the leaf vertex l, assuming the partial ordering is applied. Specifically, the cost function for a subsequence $\langle Ord_i, \ldots, Ord_{|C|}\rangle$ is given as follows:

$$cost(Ord_i) = \{ \sum_{v \in C \ s.t. \ v \notin Ord} \delta(v) \} + length(Ord_i, l)$$

The first term is a lower bound to the longest path length of the remaining unordered vertices after they have been serialized. The second term $length(Ord_i, l)$ represents the longest path length from the first element Ord_i of the subsequence to the leaf l. Theorem 7.2.5 guarantees that $cost(Ord_i)$ is always a lower bound to $length(r, l)$ in the serialized graph. The branch-and-bound strategy terminates when the first valid ordering is found.

7.4 Summary of Conflict Resolution

This section presented algorithms to resolves the conflicts that arise when multiple operations share the same hardware resource. In traditional synthesis approaches that consider only operations with bounded execution delays, conflict resolution is defined as part of the scheduling and resource binding tasks. These approaches are inappropriate for the sequencing graph model because of its support for data-dependent delay operations.

This chapter presented a conflict resolution strategy based on serializing operations in the graph model that are bound to the same resource. This strategy is based on the constraint graph model that is derived from the sequencing graph under timing constraints. In addition to incorporating data-dependent delay operations, this strategy has the following features:

- *Detailed timing constraints.* Resource conflicts are resolved subject to the required timing constraints.

- *Constraint topology.* The formulation uses the topology of the timing constraints to decompose operations into clusters. Serialization is performed at two levels: among operations within each cluster and among the clusters themselves. This strategy reduces the computational complexity of the resolution algorithms.

Relative scheduling provides the framework for analyzing consistency of timing constraints, which is used extensively in the conflict resolution formulation.

Chapter 8

Relative Control Generation

This chapter presents the synthesis of synchronous control logic to activate the functional units in the data-path according to a given schedule. Control synthesis is important because it affects the control flow of operations and hence directly impacts the overall performance of the resulting hardware. There are many different styles of control implementation, ranging from ROM-based microprogrammed controllers [SLP88] to finite-state machines [CR89] to distributed control [BCM+88]. In the simple case where the hardware model either does not contain any data-dependent delay operations or does not support multiple threads of concurrent execution flow, the control logic can be implemented using a microprogrammed controller or a single finite-state machine in a straightforward way.

In the general case, however, the hardware model may contain multiple concurrent execution sequences, each of which may depend on independent external synchronization conditions. Consider for example two execution sequences: one sequence activates a set of operations after detecting the rising edge of a signal and the other sequence depends on detecting the rising edge of another signal. Control for each execution sequence can be implemented as a finite-state machine and the overall control can be obtained by collapsing these finite-state machines together. However, this collapsing may produce a large number of states because of the independent synchronization conditions. Therefore, a systematic way of synthesizing control logic must both handle synchronization and exploit the control-flow structure to avoid computational bottlenecks. This chapter considers control generation for the general case.

We review briefly the synthesis process leading up to control generation;

the synthesis flow is illustrated in Figure 6.1. Given a sequencing graph model G_s and a resource binding β obtained through design space exploration, Hebe synthesizes a logic-level implementation of both control and data-path. Whereas the data-path structure is fixed for a given resource binding, as described in Section 4.3, synthesizing the control logic requires the following steps. First, resource conflicts in the binding are resolved by serializing operations bound to the same resource, as described in Chapter 7. Once the conflicts have been resolved, relative scheduling is performed to compute the minimum start time of each operation subject to the required timing constraints, as described in Chapter 6. Optionally, optimization techniques presented in Chapter 9 can be performed on the constraint graph to minimize the control implementation cost. Finally, control logic to activate functional units in the data-path according to this relative schedule is generated.

A novel approach called *relative control synthesis* was developed in this research to generate control logic based on relative scheduling and the hierarchical sequencing graph model with timing constraints. In this approach, the control logic is synthesized as an interconnection of interacting finite-state machines. In addition to being able to support concurrency and data-dependent delay operations in the hardware model, relative control guarantees that the resulting control implementation is *precise*, i.e., it yields the minimum number of clock cycles in executing the behavior for all input data sequences. This feature means that the control unit incurs no performance penalty for the arbitrary nesting of calls, conditionals, and loops in the input description.

Precise control implementation. As background for defining a precise control implementation, we describe first our model of the control implementation for the sequencing graph model. We assume a synchronous implementation of control that can be modeled on the whole as a synchronous finite-state machine (FSM) where transitions occur by the assertion of a clock signal at every *cycle*. The model of synchronous control as a FSM serves as an abstraction to reason about its properties; in particular, it does not imply its physical realization in hardware, i.e. the control circuit may be physically implemented either as a single FSM or as a network of FSMs.

Each graph in the sequencing graph hierarchy is associated with an abstract control FSM. For a given graph G_s, the corresponding control FSM encapsulates the control FSMs of all descendent graphs in its control-flow hierarchy. For example, consider a procedure with a loop and a loop body; the graph hierarchy consists of two graphs: the root graph for the procedure and the graph for the loop body. The resulting control also consists of two control FSMs, one for each of the two graphs.

A control FSM is characterized by a set of states called *control states*. The FSM corresponding to a control implementation of G_s is assumed to be initially in the *reset state*. Because the implementation is assumed to be non-pipelined, the FSM returns to the reset state after the last operation in the graph has completed execution. For a given input sequence, the *control delay* of the control implementation of G_s is the number of cycles to go from the reset state back to itself.

Let G denote the constraint graph that is derived from the sequencing graph G_s and detailed timing constraints (the derivation was presented in Section 4.2). The sequencing graph constrains the assignment of operations to FSM control states. Namely, no two state vertices (i.e. operations with non-zero execution delay) connected by a path in G_s can be assigned to the same control state. Equivalently, we can also say that two vertices connected by a path in the constraint graph G with strictly positive length cannot be assigned to the same control state.

More specifically, given a constraint graph G and a relative schedule $\Omega(G)$, we define a precise control implementation for G as follows.

Definition 8.0.1 *Given a constraint graph* G *and a relative schedule* $\Omega(G)$, *a control implementation for* G *is* **precise w.r.t.** $\Omega(G)$ *if its control delay is exactly equal to the difference between the sink and source start times,* $T(v_n) - T(v_0)$, *computed using the schedule* $\Omega(G)$, *for all input sequences.*

An important characteristic of preciseness is that it is defined with respect to a given schedule. For a constraint graph, there can exist multiple schedules, each of which may be used to define a precise control implementation. If the schedule is *minimum*, then the corresponding precise control implementation will yield the minimum control delay. In this case, we omit for brevity the reference to the schedule, and simply call a precise control implementation as one that is precise w.r.t. the minimum schedule.

Intuitively, the time to execute a precise control implementation of hardware behavior depends solely on the execution of the operations and not on the transfer of control. This means that the time required by the control to activate a data-dependent delay operation is precisely equal to its execution delay. For example, if an extra cycle is needed to transfer control to a called procedure (as in the microcode-based implementation of [TLW+90]), then the control implementation is not precise by the above definition. On the other hand, a precise control implementation incurs no delay penalty in the use of control flow constructs such as procedure calls, conditionals, and loops in the input behavioral description.

Recall from Section 4 the definition of the *latency* of G. The latency of a constraint graph G serves as the lower bound for the control delay. Therefore, it follows that a control implementation for G is precise w.r.t. $\Omega_{min}(G)$ if its control delay is exactly equal to the latency of G for all input sequences.

Control generation strategy. Control generation is based on an approach called *relative control synthesis*. The input to relative control consists of a hierarchical sequencing graph under timing constraints and its corresponding satisfying relative schedule; it produces as output a control unit that activates the operations according to the schedule. The control unit is implemented as an interconnection of interacting finite-state machines.

For ease of exposition, we present relative control synthesis in two parts. First, we describe a control generation approach called *adaptive* control that supports the hierarchical sequencing graph model *without* detailed timing constraints. In adaptive control, a sequencing graph is directly mapped to control logic by considering only the sequencing dependencies. It supports hierarchy by generating the proper control signals for transferring control *across* the sequencing graph hierarchy, such as in the case of procedure calls, conditionals, or loops.

When there are no detailed timing constraints, adaptive control is sufficient to generate the control unit. Relative control extends this basic technique to support detailed timing constraints by generating, for a given graph G_i, the control logic based on a given schedule $\Omega(G_i)$. As timing constraints can be applied only between operations within the same graph, the support for hierarchy is identical in the two approaches.

For both approaches, we prove that they yield a *precise* control implementation. We prove first that the adaptive control implementation is precise w.r.t. the minimum schedule in the absence of detailed timing constraints; in other words, an operation is activated as soon as all its predecessors have completed execution. We then prove that the relative control implementation is precise w.r.t. a given schedule, regardless of whether the schedule is minimal or non-minimal.

Adaptive and relative control approaches are presented in Sections 8.1 and 8.2, respectively.

8.1 Adaptive Control Approach

The *adaptive control* approach takes as input a sequencing graph model G_s without timing constraints and directly maps the graph model into a synchronous control unit consisting of a modular interconnection of interacting finite-state machines. As its name indicates, "adaptive" control takes into account the variations in the execution times of the operations caused by the changing input data.

Detailed timing constraints are not considered in the adaptive control scheme. In this case, the minimum control delay for any control implementation of G_s is simply the length of the *longest weighted path* in G_s from the source vertex v_0 to the sink vertex v_n, where the weight of a vertex is equal to its execution delay for that particular input sequence. We will show that the adaptive control implementation is *precise* by guaranteeing its control delay to be minimum for all input sequences. Extensions to support detailed timing constraints are presented in the next section.

We present an overview of the basic strategy in Section 8.1.1. Two control implementations are presented. Section 8.1.2 describes a simplified scheme that supports data-dependent delay operations and multiple execution flows, but the resulting control is not precise. We extend the simplified scheme in Section 8.1.3 to obtain a precise control implementation. Analysis of adaptive control is presented in Section 8.1.4.

8.1.1 Basic adaptive control strategy

In the case where there is a *single thread of execution flow*, a control automaton can be derived by assigning a state to each vertex of the sequencing graph [BCM+88]. State transitions in the automaton depend on the completion of the corresponding operations. The presence of multiple threads of execution flow poses a difficulty. In particular, the completion of an operation is not a sufficient condition to trigger the execution of its successor because the successor may have multiple predecessors. The activation of an operation depends on the completion of execution of all its predecessors; hence in general it is necessary to remember whether an operation has completed or not.

The adaptive control strategy addresses this difficulty by implementing control as a modular *interconnection* of control elements. There is one control element CE_i per vertex $v_i \in V$ of the sequencing graph G_s, with the interconnection of the control elements having the same topology as the sequencing graph. Since there is a one-to-one correspondence between vertices and control

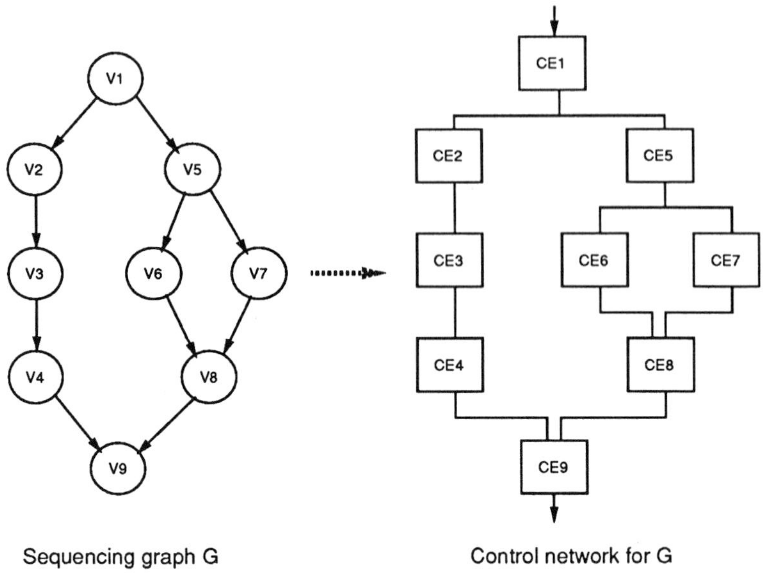

Figure 8.1: Direct mapping of the sequencing graph G to a control network.

elements, a vertex and its control element are referred to interchangeably. Figure 8.1 illustrates the mapping from a sequencing graph to a network of control elements.

Control elements coordinate with the others via handshaking signals. Two handshaking signals are defined for each control element CE_i: $enable_i$ and $done_i$. The *enable/done* signals indicate when a control element is enabled and when it has finished. In addition, communication is necessary between CE_i and functional units in the data-path. This communication takes place through an $activate_i$ signal that initiates the corresponding operation v_i in the data-path, and a $complete_i$ signal from the data-path that acknowledges the completion of operation v_i. Figure 8.2 illustrates the control element with its handshaking signals and shows the result of applying this control element model to the network of Figure 8.1.

Implementing control elements

The implementation of a control element is dependent on whether the corresponding vertex is *stateless* or *state*. When a vertex v_i is stateless for all input sequences (e.g. no-op or a combinational logic operation), its control element

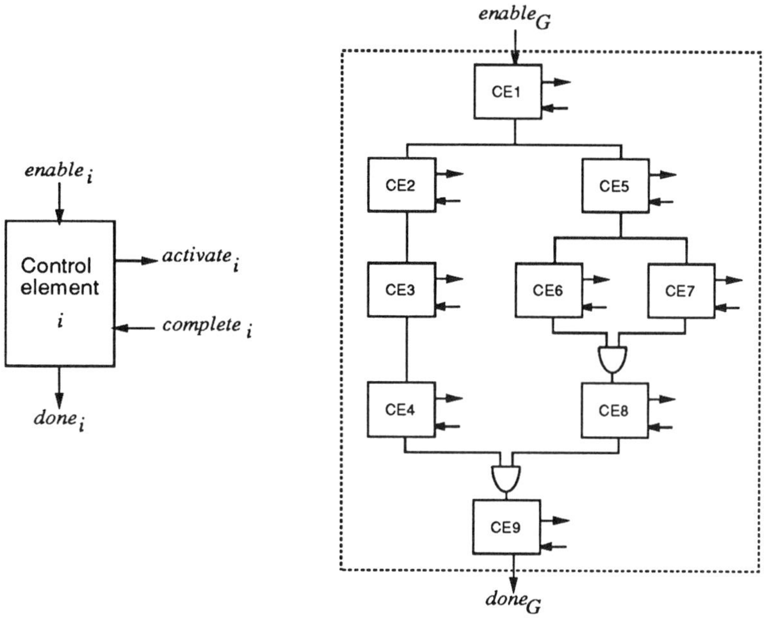

Figure 8.2: Handshaking signals for a control element and the interconnection of control elements to form a control network.

CE_i asserts the done signal as soon as it is enabled. No state information is needed in this case. The control element degenerates to combinational logic: $done_i = enable_i$.

On the other hand, when a vertex v_i requires one or more cycles of execution delay for some input sequences, its control element CE_i is implemented as a FSM with two states: a *ready* state (S_i^r) and a *wait* state (S_i^w). The initial state for a control element is the *ready* state. The reset state for an entire control network is when all control elements are in their *ready* states. In the ready state, a control element begins executing its operation whenever it is enabled. It remains in S_i^r until the completion of execution – signified by the assertion of the $complete_i$ signal – whereupon it makes a transition to the wait state. Once in the wait state, the control element cannot be activated even if it is enabled. The wait state indicates that execution has completed and the control element is waiting to be reset.

The requirement that a control element can initiate execution only in the *ready* state ensures that no race condition arises as we execute a sequence of

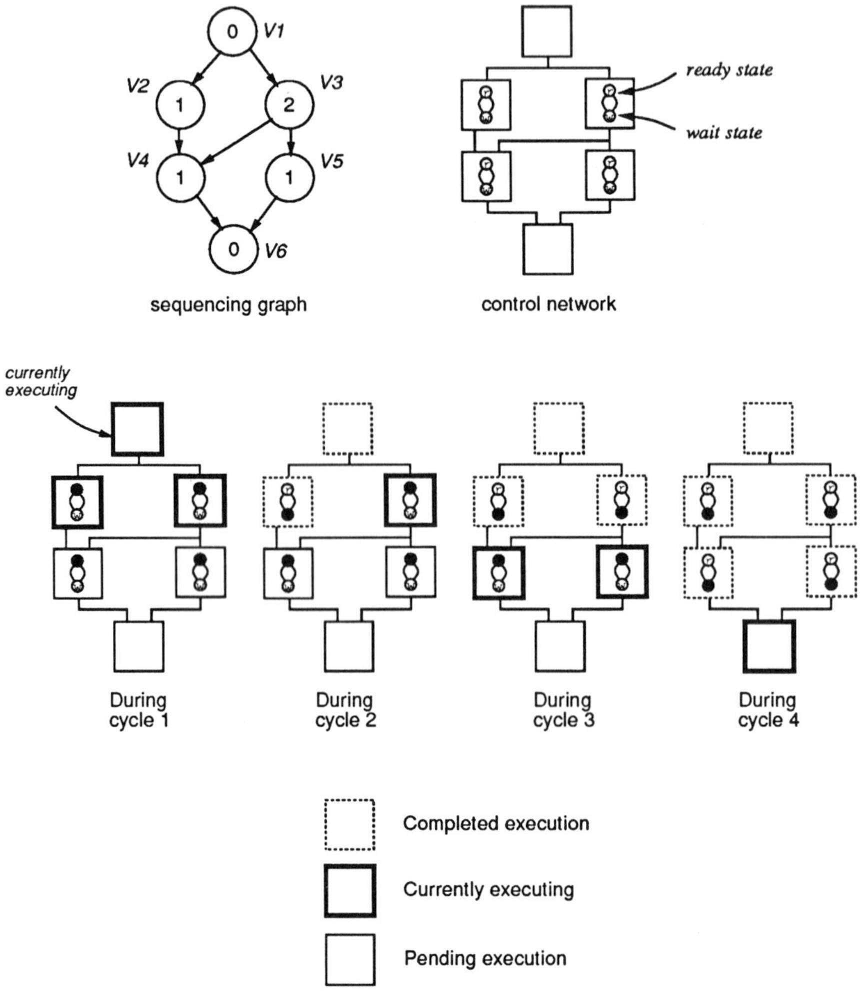

Figure 8.3: A sequencing graph with execution delays, the control network, and cycle-per-cycle execution flow.

operations. For the sake of visualizing the control mechanisms, consider the situation when the control network is reset. Execution ripples from the source to the sink. A control element upon completion makes a transition to its wait state. When the entire sequencing graph has completed execution, the network resets by having all the control elements make an uniform transition back to their ready state. Note that each control element can be activated *only once* for each execution of the sequencing graph; this strategy is based on our assumption of non-pipelined designs.

Consider for example the sequencing graph of Figure 8.3. The number in a vertex represents its execution delay. In the control network, each state vertex is represented by a two-state FSM in the corresponding control element. The execution flow of the control network is shown, where control elements in bold are currently executing their operations. The current state in the control FSMs is shaded. During the first cycle, the first two state vertices v_2 and v_3 are executing. During the next cycle, v_2 completes execution and v_3 continues execution. During the 3rd cycle, v_3 completes, which initiates the execution of v_4 and v_5. Finally, during the 4th cycle, all operations have completed execution and the control network is ready to be resetted.

The control network adapts to the changing execution delays of the operations. It has several advantages that include modularity, distribution of control, uniform handling of both fixed and data-dependent delay operations, and support for multiple concurrent execution flows. We describe now a simple adaptive control implementation that satisfies these requirements. Although it may not be precise in terms of control delay, we use this simple model to justify a more elaborate control scheme that is presented in the next section which satisfies the preciseness requirement.

8.1.2 Simple adaptive control implementation

Let $pred(v_i)$ and $succ(v_i)$ denote the set of predecessors and successors of a vertex v_i in the sequencing graph G_s, respectively. We define the handshaking signals for a control element CE_i below:

$$enable_i \equiv \prod_{k \in pred(v)} done_k \tag{8.1}$$

$$activate_i \equiv \begin{cases} S_i^r \cdot enable_i & v_i \text{ state} \\ 0 & v_i \text{ stateless} \end{cases} \tag{8.2}$$

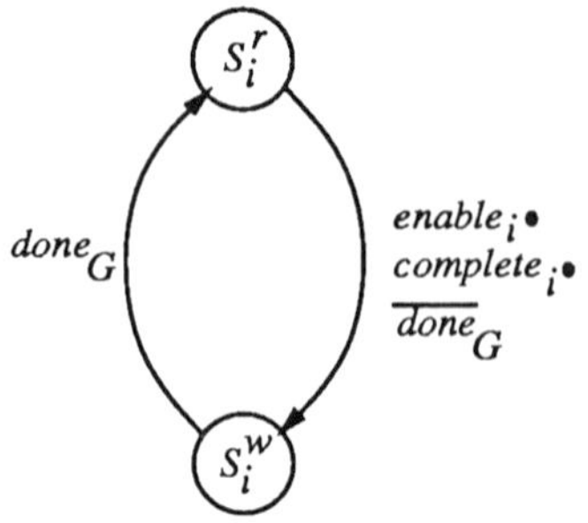

Figure 8.4: State transition diagram for the simple control element.

$$done_i \equiv \begin{cases} S_i^r \cdot enable_i \cdot complete_i + S_i^w & v_i \text{ state} \\ enable_i & v_i \text{ stateless} \end{cases} \tag{8.3}$$

Note that $complete_i$ is generated by the data-path.

A vertex is enabled when all predecessors have completed execution, whereupon the corresponding operation is activated until its completion. The enable of a sequencing graph G_s is the enable of its source vertex, denoted by $enable_G$; likewise, the done of a sequencing graph G_s is the done of its sink vertex, denoted by $done_G$. The enable signal of the root graph corresponding to a *process* model in HardwareC is a constant logic "1" to ensure that the hardware always restarts execution upon completion.

When a vertex v_i is stateless for all input sequences, it asserts its done signal as soon as it is enabled. Otherwise, as described in the previous section, the control element is implemented as the two-state FSM shown in Figure 8.4. The FSM remains in the ready state (S_i^r) until it has completed execution, after which it enters the wait state. The transition back from S_i^w occurs when the entire graph G_s has completed execution, signaled by the assertion of $done_G$. The transition conditions for the FSM are given below:

$$S_i^r \longrightarrow S_i^w \qquad enable_i \cdot complete_i \cdot \overline{done_G}$$

$$S_i^w \longrightarrow S_i^r \qquad done_G$$

Note that upon completion, a vertex remains in the ready state if $done_G$ is also asserted. This corresponds to the case when the completion of this vertex results in the completion of the entire sequencing graph.

Support for hierarchy

The control element model is applicable to any type of vertices. The personalization is determined by how the control elements interconnect their *activate* and *complete* signals. For *simple* vertices, these signals are connected directly to the data-path components implementing the operations. For instance, the activate/complete signals for input/output vertices are connected to the I/O ports. For load-register vertices, they are connected to the registers being loaded. For message passing send/receive vertices, the activate/complete signals are mutually coupled. Specifically, the activate of the send is connected to the complete of the corresponding receive, and the activate of the receive is connected to the complete of the send.

Additional control circuitry is needed for *complex* vertices. They are illustrated in Figure 8.5 and are described below:

- **Call Vertex** – A network of control elements implementing a sequencing graph can be treated as a single *abstract* control element, where the *enable/done* of this abstract control element is connected to the enable/done of the given graph. This formulation supports hierarchy; it also allows a consistent view of the control for both a single operation and a group of operations. The *activate* (or *complete*) signal of the call is connected to the enable (or done) of called sequencing graph, as shown below.

$$enable_H \equiv activate_i \quad (8.4)$$

$$complete_i \equiv done_H \quad (8.5)$$

  where $enable_H$ and $done_H$ are the enable and done of the called graph H, respectively.

- **Conditional Vertex** – Since each branch of a conditional is a separate sequencing graph, a conditional is analogous to a multi-way procedure call that depends on the value of a conditional expression $cond$. For a conditional with k branches, the handshake signals are defined as follows:

$$enable_{G(j)} \equiv activate_i \cdot (cond = j^{th} branchvalue) \quad (8.6)$$

$$complete_i \equiv \sum_{n=1}^{k} done_{G(j)} \quad (8.7)$$

  where $enable_{G(j)}$ and $done_{G(j)}$ represent the enable and done of the sequencing graph $G(j)$ corresponding to the jth branch, and $activate_i$ and

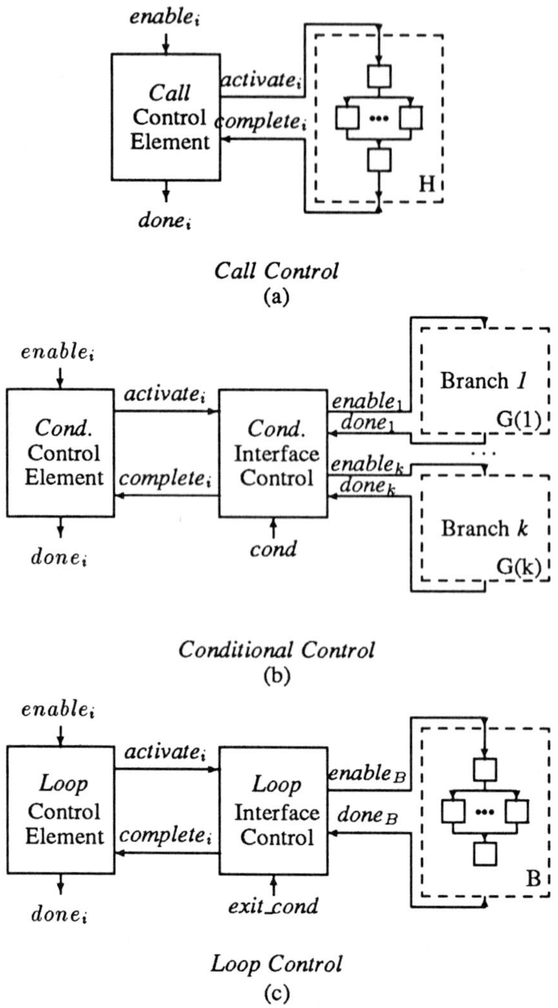

Figure 8.5: Control for complex vertices: (a) call, (b) conditional, (c) loop.

$complete_i$ represent the activate and complete signals of the conditional vertex. One and only one branch is selected at any particular time, hence the completion of a conditional can be computed as the disjunction of the done signals of the branches. We assume here that the branch condition remains unchanged for the entire duration of execution of the conditional vertex. If this assumption is not valid, then either the value of the conditional expression or the activate signals $enable_{G(j)}$ must be stored to ensure that the branch condition remains unchanged during execution of the conditional.

- **Loop Vertex** – A loop executes its body until the exit condition ($exit\_cond$) becomes true. The generation of the exit condition depends on whether the loop bounds are fixed or data-dependent. For fixed iteration loops, a straightforward approach is to use a counter that steps through each iteration of the bounded loop. For the more general case of data-dependent loops, $exit\_cond$ is an expression whose value may change dynamically. In particular, exit from a `repeat-until` loop takes place when both $exit\_cond$ is true *and* the loop body has finished execution, as shown below.

$$enable_B \equiv activate_i \tag{8.8}$$

$$complete_i \equiv exit\_cond \cdot done_B \tag{8.9}$$

where $enable_B$ and $done_B$ are the enable and done of the loop body B, and $activate_i$ and $complete_i$ correspond to the loop vertex. Note that *while* loops are assumed to be transformed to conditional *repeat-until* loops, so that in the sequel all loops are assumed to be repeat-until loops.

8.1.3 Precise adaptive control implementation

A limitation of the simple control implementation described in the previous section is that it may not be precise. An inherent difficulty lies in the fact that *resetting* of the control network occurs *too late in time*. There is a conflict between trying to reset the network in preparation for the next activation and actually knowing when to reset.

To illustrate the difficulty, consider a sequencing graph G_s representing either a process or a loop body that we wish to execute repeatedly. At a particular cycle n, G_s is in its final cycle of execution and $done_G$ is asserted at the start of the next cycle $n + 1$. Ideally, the control of G_s should be ready to restart execution during cycle $n + 1$. However, this is not possible in the simple control implementation because resetting of the control network occurs one cycle after the assertion of $done_G$. Consider the example in Figure 8.3. The control network detects the completion of the entire graph during cycle 4, and resets one cycle later in cycle 5. Therefore, the periodicity of execution is 4 cycles instead of the critical path delay (latency) of 3 cycles. This is clearly non-minimal in terms of control delay, and hence it is not a precise control implementation.

When all operations have fixed delays, the restarting periodicity can be hard-coded into the control because the latency of the graph is fixed. We say in this case that the control equations can be *statically* derived. Conversely, when data-dependent delay operations are present, the control equations have to consider *dynamically* variations in the input signals. Since the latency of the graph may change, the hard-coded control approach cannot be used in general. To resolve this difficulty, two mechanisms are used to construct a precise control implementation. The first mechanism is to use *lookahead* to ensure proper resetting of the control for all input sequences. The second mechanism is to *dynamically* identify *stateless* operations.

Lookahead for Resetting

To ensure proper and timely reset of the control elements, it is necessary to know when a given graph is in the *last cycle* of its execution. If the detection of completion occurs during the last cycle, then the control can reset during the last cycle and restart execution in the following cycle. This guarantees that the control of a sequencing graph can repeatedly execute without losing a cycle to reset, which is crucial to ensure no wasted time for restarting loops and processes.

This look-ahead requirement can be achieved by defining the *complete* signal of a vertex to be asserted during the last cycle of execution of the corresponding operation. For example, suppose an operation takes only one cycle, the corresponding complete signal is asserted as soon as it is enabled. In general, if an operation takes n cycles to execute, the corresponding complete signal will be asserted *during* the nth cycle. Note that it is straightforward to implement a control scheme that uses this look-ahead technique. The only complication is the necessity to determine whether the loop exit condition is satisfied *during* the final cycle of a loop's execution. Since the exit condition may depend on values of variables that are stored in registers and updated at the end of each iteration, the exit condition is computed with the values of variables before they are updated into the registers.

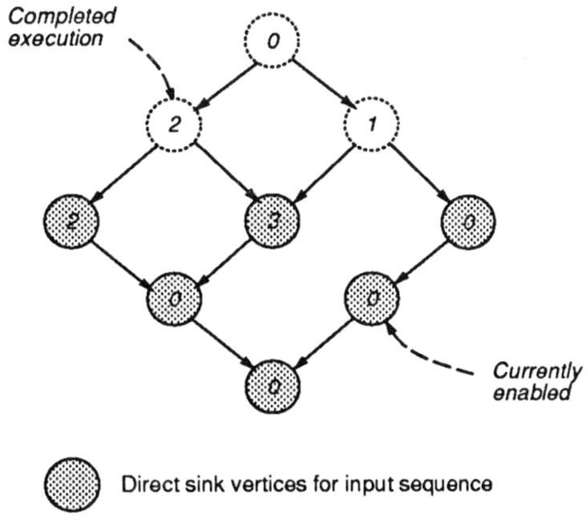

Figure 8.6: Example of direct-sink vertices, shown as shaded vertices.

The redefinition of the complete signals needs to be applied consistently throughout the sequencing graph model. Whereas the redefinition is straightforward for simple vertices, it poses a difficulty for complex vertices. In particular, for complex vertices that invoke other sequencing graphs, we need to detect the completion of an entire graph during the final cycle of execution, denoted by the assertion of $done_G$. Since a sequencing graph may have multiple threads of concurrent execution flow, we identify the set of vertices, called *direct-sink* vertices, whose combined completion results in the completion of the entire graph.

Definition 8.1.1 *A vertex v of a sequencing graph graph G_s is a* **direct-sink** *vertex if the longest weighted path from v to the sink, excluding the weight of v, is zero. The weight for a vertex v is equal to its execution delay $\delta(v)$ for a particular input sequence.*

As an illustration, consider the sequencing graph of Figure 8.6, where a number in a vertex represents its execution delay *for a particular input sequence.* Shaded vertices denote direct-sink vertices. The completion of execution of all direct-sink vertices results in the completion of the execution of the entire graph. Note that if the source vertex is direct-sink for a particular input sequence, then by definition the sequencing graph is stateless for that input sequence.

A complication arises since the execution delay of a data-dependent vertex may change for different input sequences. In particular, the delay may become zero, making the vertex stateless. Therefore, it is impossible to always statically identify the set of direct-sink vertices. To address this issue, a direct-sink signal ($dsink$) is defined for each control element of the network, which is asserted when the corresponding vertex is currently direct-sink. Specifically, the $dsink_i$ signal for a vertex v_i is computed as follows:

$$dsink_i \equiv \prod_{s \in succ(v)} dsink_s \cdot stateless_s \qquad (8.10)$$

where $stateless_s$ is asserted if the corresponding vertex s is stateless. Note that $dsink$ of the sink vertex is a constant logic "1."

With this formulation, the control can track the variations in the execution delay of a graph because the $dsink$ signals are evaluated dynamically during hardware execution. When the direct-sink vertices are all bounded vertices, the $dsink$ signals can be evaluated statically, not adding to the complexity of the resulting control implementation.

Dynamically Identify Stateless Computation

Computations that do not take any time to execute, such as stateless branches of conditionals, should immediately activate their successor operations without taking a cycle to transfer control. It is therefore important to know when a given operation is stateless; i.e. execution delay is zero for a particular input sequence. Since our model supports data-dependent delay operations, it is necessary to *dynamically* determine whether a particular operation is stateless.

We define a *stateless* signal for each control element which is asserted whenever the corresponding vertex is stateless for a particular input sequence. The stateless signals for simple vertices are known statically, e.g. the stateless signal for load register, synchronous I/O, or synchronous send/receive message passing vertex is always "0". The stateless signals for complex vertices are evaluated dynamically, as described below:

- *Call* – The stateless signal for a call vertex is asserted if the called graph is stateless:

$$stateless_i \equiv dsink_H \qquad (8.11)$$

where $dsink_H$ is the direct-sink signal of the source vertex of the called graph H, which is asserted if H is stateless.

```
ALU( a, b, result, opcode )
    in boolean a[8], b[8], opcode[3];
    out boolean result[8];
{
    if ( opcode == OP_DIV ) {
        /* division requiring more than one cycle */
        result = a / b;
    } else if ( opcode == OP_AND ) {
        /* simple logic operation */
        result = a & b;
    } else
        /* invalid opcode - do nothing */
}
```

Figure 8.7: Example of ALU description with stateless and state branches.

- *Conditional* – The stateless signal is asserted whenever the selected branch is stateless:

$$stateless_i \equiv \sum_{j=1}^{k} dsink_{G(j)} \cdot (cond = j^{th} branchvalue) \quad (8.12)$$

  where $dsink_{G(j)}$ is the direct-sink signal of the source vertex of the sequencing graph $G(j)$ corresponding to the jth branch, and $cond$ is the conditional expression.

- *Loop* – For a `repeat-until` loop, the $stateless_i$ signal is always "0" because the loop body executes at least once. From Chapter 2, we note that a `while-do` loop has already been transformed into a conditional repeat-until loop.

The use of stateless signals extends the flexibility and power of the control to adapt to the input variations. Most systems deal with loops and procedure calls statically. For example, a procedure is often tagged as being either *combinational* or *sequential* at compile time, depending on whether it requires zero or more cycles to execute, respectively. However, the implementation of a procedure may be at times combinational and at times sequential, depending on the inputs that are applied.

Consider the simple description of an ALU shown above. When the hardware modeled by `ALU()` is executing, the stateless branch of the conditional is selected if `opcode == OP_AND`, whereupon `ALU` becomes a combinational procedure. Otherwise, it is a sequential procedure if the opcode selects an

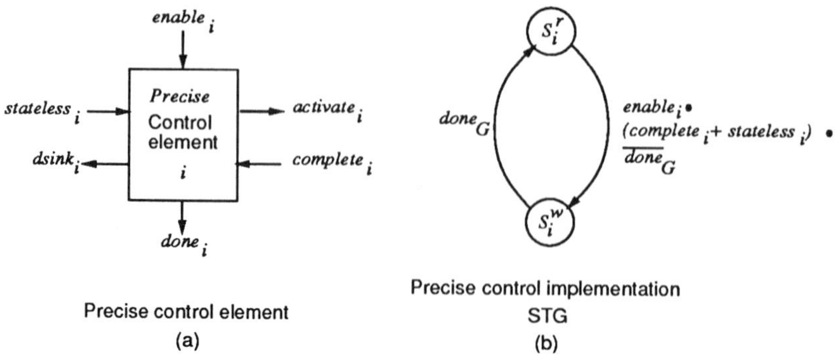

Figure 8.8: Precise adaptive control element and its STG.

operation requiring one or more cycles to execute, such as OP_DIV. By using the *stateless* signal, we can now take into account dynamic variations in the time required to call a procedure, and not waste a cycle in calling stateless (combinational) procedures.

Implementing precise adaptive control

The precise control implementation extends the approach described in the previous section by incorporating the mechanisms of look-ahead resetting and dynamically identifying stateless computations. The control element for a state vertex v_i, shown in Figure 8.8, has two states as in the previous model: *ready* S_i^r and *wait* S_i^w. The transition conditions are now described as follows:

$$S_i^r \longrightarrow S_i^w \qquad enable_i \cdot (complete_i + stateless_i) \cdot \overline{done_G}$$

$$S_i^w \longrightarrow S_i^r \qquad done_G$$

The operation begins executing in the ready state. Upon completion or detection of statelessness, the FSM makes a transition to the wait state. When the entire graph is done (signified by $done_G$), the control element makes an uniform transition back to the ready state.

The enable signal $enable_i$ of a control element CE_i remains the same as in the simple control implementation, equal to the conjunction of its predecessor's done signals (Equation 8.1). However, the $activate_i$ and $done_i$ signals are modified as follows:

$$activate_i \equiv \begin{cases} S_i^r \cdot enable_i \cdot \overline{stateless_i} & v_i \text{ state} \\ 0 & v_i \text{ stateless} \end{cases} \quad (8.13)$$

$$done_i \equiv \begin{cases} S_i^r \cdot enable_i \cdot (stateless_i + \\ \quad complete_i \cdot dsink_i) + S_i^w & v_i \text{ state} \\ enable_i & v_i \text{ stateless} \end{cases} \quad (8.14)$$

The $activate_i$ signal is asserted when the FSM is both enabled and in the ready state (S_i^r). The main difference between the implementation presented in Section 8.1.2 and the one described above is that the $complete_i$ signal is asserted *during* the last cycle of execution. Therefore, the control element can detect that its operation is currently in the last cycle and makes a transition to the *wait* state (S_i^w) for the next cycle.

8.1.4 Analysis of adaptive control

In this section, we will prove that the implementation presented in the previous section is *precise*, i.e. the control delay is equal to the latency of the sequencing graph $G_s(V, E_s)$ for all input sequences. Note that minimum and maximum timing constraints are not considered in this analysis. Therefore, the latency of G_s is the length of the longest weighted path from the source vertex to the sink vertex, where the weight of a vertex is equal to its execution delay.

We begin by decomposing the definition of preciseness into two criteria, both of which must be satisfied by a precise control implementation. We say that a control element executes when it activates the corresponding operation, i.e. asserts the activate signal, and a control element completes execution when the corresponding operation completes execution. Let $pred_{state}(v_i) \subset V$ denote the set of state vertices such that a state vertex v_p is in $pred_{state}(v_i)$ if there is a path of stateless vertices from v_p-to-v_i.

Lemma 8.1.1 *Given an input data sequence, the control delay of a synchronous control implementation of a sequencing graph G_s without timing constraints is equal to its latency if the following two criteria are satisfied:*

1. *Precise transition: For every state vertex $v_i \in V$, the control element is enabled in the cycle after the completion of execution of all state vertices in $pred_{state}(v_i)$.*

2. *Precise restarting: For all polar sequencing graphs in the hierarchy induced by G_s, the control element of the source vertex of a given graph is activated in the cycle following the completion of all its vertices.*

Proof: Given an input data sequence, the execution delays for all operations in G_s are known. Therefore, the latency of G_s can be computed as the longest weighted path in G_s. The precise transition criterion ensures that transferring control from any control element to its successors incurs no penalty in terms of cycles for state vertices. Recall that the done signals of stateless vertices are immediately asserted upon being enabled, and hence stateless vertices do not affect the execution delay and control delay.

The precise restarting criterion ensures that transferring control across the hierarchy also does not incur penalties. In particular, it ensures that: (1) the control delay for a call vertex is equal to the control delay of the call graph, (2) the control delay for a conditional vertex is equal to the control delay of the graph corresponding to the selected branch, and (3) the control delay for a loop vertex is equal to the control delay of the graph corresponding to the loop body times the number of iterations. Therefore, if both criteria are met, then for a given input data sequence, the control delay for the synchronous control implementation of G_s is equal to the latency of G_s. ||

By definition, the complete signals for simple vertices are asserted during the final cycle of execution of the corresponding operations. We show now that this is true also for the complex vertices: *call*, *conditional*, and *loop*.

Lemma 8.1.2 *In the control implementation of Section 8.1.3, the complete$_i$ signal for any complex vertex $v_i \in V$ in a sequencing graph $G_s(V, E_s)$ is asserted during the final cycle of execution of v_i.*

Proof: From Equation 8.5, the complete signal of a call vertex is equal to the done of the called graph. From Equation 8.7, the complete signal of a conditional vertex is the disjunction of the done signals from each of the conditional branches. From Equation 8.9, the complete signal of a loop vertex is the conjunction of the loop exit condition and the done of the loop body, where the loop exit condition asserts during the final cycle of execution of the loop. Therefore, it is sufficient to show that the done of a sequencing graph G_s is asserted during the final cycle of execution of G_s.

This can be shown by an inductive argument. Consider first the case where the graph G_s consists of only simple vertices. The done signal of G_s, $done_G$, is the conjunction of the done signals of all direct sink vertices. From the definition of $dsink_i$ (Equation 8.10) and $done_i$ (Equation 8.14), the done signals for all direct-sink vertices v_i are asserted as soon as their corresponding $complete_i$ signals are asserted. Since all vertices are simple, $complete_i$ is asserted during the final cycle in the execution of v_i, and $done_G$ is also asserted during the final cycle of execution of G_s.

Now consider a graph containing complex vertices, where the done signals of the graphs associated with the complex vertices are asserted during their respective final cycle of execution. By Equations 8.5, 8.7, and 8.9, the complete signals of the complex vertices are also asserted during the final cycle of execution. From the above argument, the done signal of the graph containing the complex vertices is also asserted in the final cycle of its execution. ||

Lemma 8.1.3 *The precise adaptive control implementation of G_s satisfies the precise transition criterion.*

Proof: The precise transition criterion states that for the control element of a state vertex v_i is enabled in the cycle after all the assertion of done signals from every control element in $pred_{state}(v_i)$. From Equation 8.1, a control element is enabled when the done signals of all its predecessors are asserted, e.g. $enable_i = \prod_{v_p \in pred(v_i)} done_p$. If v_p is a stateless vertex, then it asserts its done signal as soon as it is enabled, e.g. $done_p = enable_p$. Therefore, the $enable_i$ signal for v_i is asserted when the $done_p$ signals of all $v_p \in pred_{state}(v_i)$ are asserted, i.e. $enable_i = \prod_{v_p \in pred_{state}(v_i)} done_p$

To prove the precise transition criterion, it is sufficient to show that for all vertices $v_p \in pred_{state}(v_i)$, the $done_p$ signal is asserted in the cycle after the completion of v_p. This implies that $enable_i$ is also asserted in the cycle after the completion of all state vertices $v_p \in pred_{state}(v_i)$.

Note that since v_i is a state vertex, any vertex $v_p \in pred_{state}(v_i)$ is by definition indirect-sink vertex, e.g. $stateless_p = dsink_p = 0$. Therefore, Equation 8.14 reduces to $done_p = S_p^w$. Since v_p enters the wait state S_p^w in the cycle after the assertion of $complete_p$, and since by Lemma 8.1.2 $complete_p$ is asserted during the final cycle of execution of v_p, the signal $done_p$ is asserted in the cycle after the completion of v_p. This holds for every $v_p \in pred_{state}(v_i)$, and hence the precise transition criterion is satisfied. ||

Lemma 8.1.4 *The precise adaptive control implementation of G_s satisfies the precise restarting criterion.*

Proof: From Lemma 8.1.2, the done of a graph G_s is asserted during the final cycle of its execution, at which time all control elements make a uniform transition back to the ready states. Since a vertex can begin execution when enabled in the ready state (Equation 8.13), during the next cycle the source vertex can immediately begin activation if it is enabled. Therefore, the control network is ready to restart execution in the cycle after the sink completes execution. ||

We now prove the adaptive control implementation in the previous section to be precise.

Theorem 8.1.1 *The precise adaptive control implementation of G_s without timing constraints is precise.*

Proof: For a given input sequence, the adaptive control implementation in the previous section satisfies both the precise transition (Lemma 8.1.3) and precise restarting (Lemma 8.1.4) criteria. By Lemma 8.1.1, the control delay is equal to the latency of the sequencing graph for a given input sequence.

The categorization of vertices to stateless versus state and direct-sink versus indirect-sink is dynamically evaluated through the use of the *stateless* and *dsink* signals for each control element. Therefore, Lemma 8.1.3 and Lemma 8.1.4 hold for all input sequences. Since Lemmas 8.1.1 is satisfied for all input sequences, the adaptive control implementation is precise. ||

8.2 Relative Control Approach

Although adaptive control is a general and flexible control synthesis technique for the sequencing graph model, the method is limited in that timing constraints are not considered. Since timing constraints are incorporated in the relative scheduling formulation, we introduce *relative control* as an extension of adaptive control to support control generation from a relative schedule.

Consider a constraint graph $G(V, E)$ that is derived from a sequencing graph G_s under timing constraints. A relative schedule of G, denoted by $\Omega(G)$, defines for each operation $v \in V$ a start time $T(v)$ in terms of time offsets $\sigma_a(v)$ from the completion of a set of anchors $a \in A(v)$. As in adaptive control, the resulting control is implemented as an interconnection of finite state machines. The support for hierarchy, i.e. call, conditional and loop vertices, remains identical to that presented in Section 8.1.3. However, the generation of the *enable* and *done* handshaking signals is different. In particular, instead of activating an operation v as soon as its predecessors have completed execution, as is the case for adaptive control, we now activate an operation based on the condition specified by its start time derived from relative scheduling.

Intuitively, since the activation of an operation is defined with respect to the anchors, we can divide the control synthesis task into two steps – control generation for the non-anchors, and control generation for the anchors.

- *Generating control for non-anchors.* The control logic to activate fixed delay operations (non-anchors) is first generated under the assumption that the completion of each anchor $a \in A$ is indicated by the assertion of a signal $done_a$.

- *Generating control for anchors.* The control logic to activate the anchors, as well as detecting the completion of anchors, is synthesized.

Details of each step are presented in the following sections.

8.2.1 Relative control for non-anchors

This section describes the generation of the $enable_i$ and $done_i$ signals for the fixed delay operations $v_i \in \{V - A\}$ (non-anchors). We assume the completion of an anchor $a \in A$ is indicated by the assertion of a signal $done_a$, which stays asserted until all operations in the graph have completed execution. The generation of $done_a$ for the anchors is described in Section 8.2.2.

Before describing the general formulation of control, we illustrate two possible alternatives for control generation in Figure 8.9. The figure shows the circuit that enables an operation v. For the sake of illustration, we assume the anchor set of v comprises of a and b, with offsets $\sigma_a(v) = 2$ and $\sigma_b(v) = 3$, respectively. This implies that v must start at least 2(3) cycles after the completion of operation $a(b)$. The two schemes are described below.

- *Counter based control* – The most intuitive and direct approach is to use a counter for each anchor that starts to count upon the completion of the anchor. The enable signals can then be described as comparisons between the values of the counters with the corresponding offsets. Let $Counter_a$ denote the value of the counter for anchor a. The enable signal is defined as follows.

$$enable_v = \prod_{a \in A(v)} \{Counter_a \geq \sigma_a(v)\}$$

  where the $\prod$ in the expression denotes logical conjunction. If the execution delay of v ($\delta(v)$) is fixed, then the done signal can also be defined in

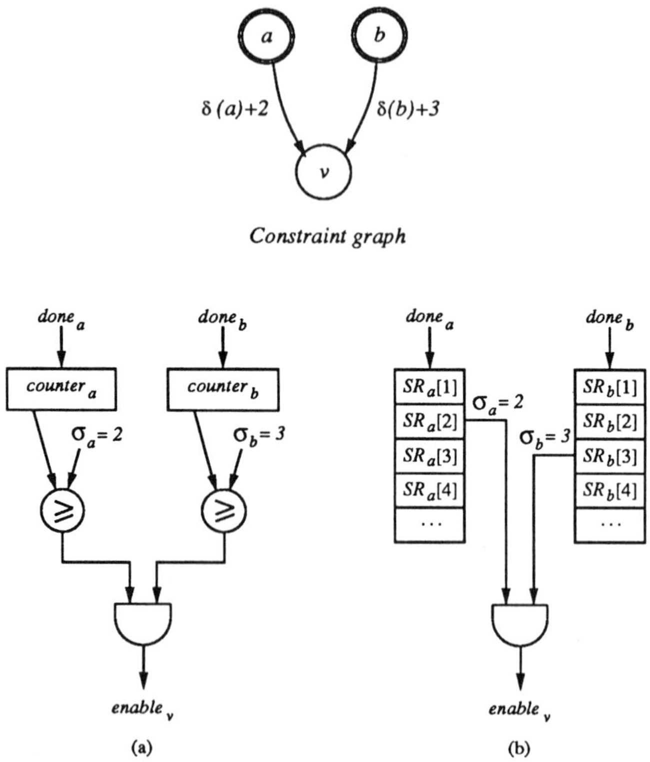

Figure 8.9: Alternate implementation styles for control generation: (a) counter-based and (b) shift-register based.

an analogous manner.

$$done_v = \prod_{a \in A(v)} \{Counter_a \geq (\sigma_a(v) + \delta(v))\}$$

The activate signal in this case is defined as $activate_v = enable_v \cdot \overline{done_v}$. No completion signal ($complete_v$) is required for fixed delay operations because $\delta(v)$ is known *a priori*.

If there is a single irredundant anchor in the anchor set $A(v)$, then we can simplify the generation of $activate_v$ by directly asserting it based on specific counter values. For example, if the delay of v is 2 and there is a single anchor $a \in A(v)$, then the activate signal is: $activate_v = (Counter_a \equiv 2) + (Counter_a \equiv 3)$, a significant reduction in complexity.

Although the control scheme is straightforward, the amount of combinational logic that is needed for the comparisons may be large. In the simple case without data-dependent delay operations, the control reduces to a single counter, which can be implemented using either microprogrammed controllers or state machines.

- *Shift-register based* – The comparator cost in the previous approach can be reduced by using shift registers instead of counters. For an anchor $a \in A$, we create a shift register SR_a whose input is the $done_a$ signal corresponding to the completion of the anchor a. The output of stage i of the shift register is denoted by $SR_a[i]$, and is asserted when *at least* i clock cycles have elapsed since the completion of a. With this formulation, the enable signal for an operation v is defined as follows.

$$enable_v = \prod_{a \in A(v)} SR_a[\sigma_a(v)]$$

Likewise, if v is a fixed delay operation, then its done signal is defined as follows.

$$done_v = \prod_{a \in A(v)} SR_a[\sigma_a(v) + \delta(v)]$$

Note the significant savings in terms of combinational logic compared with the previous approach at the expense of higher register costs. We see that tradeoffs can be made between register and combinational logic in the control implementation. The final decision rests on both the cost parameters of the logic elements, as well as on the resulting schedule.

We use the full anchor set $A(v)$ for control generation in this section. The extension to use the irredundant anchor set $IR(v)$ is straightforward and can be achieved by replacing $A(v)$ with $IR(v)$ for each vertex $v \in V$.

Remarks. We can make two important observations from the control schemes described in the previous section. The first observation is that the control logic to activate the non-anchors can be divided into the two following portions:

- *Offset control circuit* for each anchor $a \in A$. The offset control circuit generates signals that indicate the time offset from the completion of a given anchor.

- *Synchronization control circuit* for each vertex $v \in V$. The synchronization control circuit coordinates the activation of an operation with respect to offsets in its start time.

For the counter-based scheme, the offset control for an anchor corresponds to a counter; the synchronization control for a vertex corresponds to a comparator and a logical-*and* gate. For the shift register-based scheme, the offset control corresponds to a shift register and the synchronization control corresponds to a logical-*and* gate. A way to visualize the interaction between offset and synchronization control is to view each anchor as a distinct "schedule"; the activation of an operation may depend on multiple schedules. The schedule is the offset control and the logic to coordinate the activation of the operation is the synchronization control.

The second observation is that the *size* of the offset control for an anchor a is equal to the maximum offset among all vertices in the graph with respect to a. We define the following.

$$\sigma_a^{max} \equiv \max_{\forall v \in V \ s.t. \ a \in A(v)} \left\{ \begin{array}{ll} \sigma_a(v) & \text{if } v \text{ anchor} \\ \sigma_a(v) + \delta(v) & \text{if } v \text{ non-anchor} \end{array} \right\} \quad (8.15)$$

The value σ_a^{max} represents the largest offset that is referenced with respect to anchor a in the graph.

Abstracting the control implementation. With the partitioning of the control into offset and synchronization control portions, we can abstract the control implementation in terms of a state transition diagram (STG). By varying the state encoding of the STG, different control implementations will result, each corresponds to a particular tradeoff between register and logic literal costs.

The state transition diagram is formulated as a sequencer that starts stepping through the states upon assertion of $done_a$, as shown in Figure 8.10. The states of the STG are $\{s0, s1, \ldots, sk\}$, where $k = \sigma_a^{max}$. Initially, the STG is in $s0$. When $done_a$ is asserted, the states are sequenced until $reset$ is asserted, whereupon the STG returns to the initial state $s0$. Since the control FSM resets when the entire graph has completed execution, its reset signal is connected as follows: $reset = done_G$.

The output of the offset control consists of the signals $C_a(i), i = 0, \ldots, \sigma_a^{max}$, where the assertion of $C_a(j)$ means that *at least* j cycles have elapsed since the completion of anchor a. Based on this model of offset control, the synchronization control logic for a vertex v can be implemented as a logical conjunction of the respective offset control signals. Specifically, the enable, done, and activate signals for a fixed delay vertex v is:

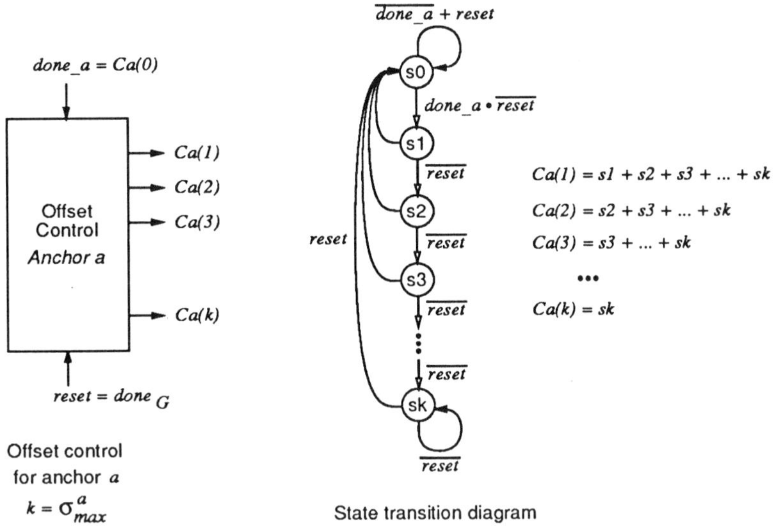

Figure 8.10: Modeling the offset control for an anchor a: (a) block diagram, (b) state transition diagram.

$$enable_v = \prod_{a \in A(v)} C_a(\sigma_a(v)) \tag{8.16}$$

$$done_v = \prod_{a \in A(v)} C_a(\sigma_a(v) + \delta(v)) \tag{8.17}$$

$$activate_v = enable_v \cdot \overline{done_v} \tag{8.18}$$

It should be evident from the analysis of the importance of removing redundant anchors for control generation. The savings arise in two areas. First, it reduces the cost of the synchronization logic in generating the *enable* signals. Second, it reduces the maximum offset σ_a^{max} corresponding to an anchor $a \in A$ by taking advantage of the cascading effect of anchors and hence reduces the number of registers in the control implementation.

8.2.2 Relative control for anchors

The enable signals are generated via the offset control and synchronization control strategy described in the previous section. The task that remains is to generate the control logic to activate the anchors.

A strategy similar to the precise adaptive control implementation is used to synthesize for each anchor $a \in A$ a control element consisting of a two-state FSM (Figure 8.8). The enable signal for an anchor $a \in A$ is generated in the same fashion as for non-anchors, i.e.

$$enable_a \quad = \quad \prod_{q \in A(a)} C_q(\sigma_q(a)) \tag{8.19}$$

where we assume the existence of the offset control model of the previous section. The $activate_a$ and $done_a$ signals for the anchor are also generated by the corresponding control FSM.

To ensure that the control implementation is precise, we perform the following.

- *Lookahead for resetting.* The $complete_a$ for an anchor $a \in A$ is defined to be asserted during the final cycle of execution. This involves two conditions. First, $done_G$ must be asserted during the final cycle of its execution. In particular, $done_G$ is defined to be asserted one cycle before the activation of the sink vertex v_n.

  $$done_G \quad = \quad \prod_{q \in A(v_n)} C_q(\max\{0, \sigma_q(v_n) - 1\}) \tag{8.20}$$

  In other words, if the offset $\sigma_q(v_n)$ is zero, then $done_G$ depends on $C_q(0)$; otherwise, $done_G$ depends on time offset one cycle before the activation of v_n. The complete signals for call, conditional and loop vertices remain the same as in adaptive control.

  Second, it is necessary to dynamically determine whether an anchor is *direct-sink* or not. Let $\widehat{A} = A \cup \{v_n\}$ be the set of anchors A augmented with the sink vertex v_n and let $\widehat{Q}(a) = \{q | q \in \widehat{A} \vee \sigma_a(q) = 0\}$ be the successors to an anchor a for which the corresponding offset is zero. The direct-sink signal $dsink_a$ for an anchor $a \in A$ is defined as follows.

  $$dsink_a \quad \equiv \quad \prod_{q \in \widehat{Q}(a)} dsink_q \cdot stateless_q \tag{8.21}$$

  where $stateless_{v_n} = dsink_{v_n} = 1$. Therefore, $dsink_a$ is asserted if all elements $q \in \widehat{A}$ are stateless, which means that the completion of a may result in the completion of the entire graph.

- *Dynamically identify stateless computation.* The $stateless_a$ for an anchor $a \in A$ is defined to be asserted whenever the corresponding operation is stateless. For simple vertices, the stateless signals are known statically. For complex vertices, these signals are defined as in adaptive control.

8.2.3 Analysis of relative control

In this section, we will prove that the relative control implementation is precise with respect to a given relative schedule. Consider a constraint graph G that is derived from a sequencing graph G_s under timing constraints, and a relative schedule $\Omega(G)$, we will prove that the difference between $T(v_n)$ and $T(v_0)$ is equal to the control delay, thereby proving the relative control implementation to be precise.

Theorem 8.2.1 *For a constraint graph G derived from a sequencing graph G_s with timing constraints, and a relative schedule $\Omega(G)$, the relative control implementation is precise w.r.t. the schedule $\Omega(G)$.*

Proof: The proof is analogous to the proof for Theorem 8.1.1. Specifically, we need to show that for a given input sequence, the *precise transition* and *precise restarting* criteria are satisfied. If this is true, then since the properties of stateless and direct-sink are dynamically determined, the control implementation is precise.

We show first that the precise transition criterion is satisfied. By the definition of a valid schedule $\Omega(G)$, the time offset from an anchor a to an operation v satisfies the timing and sequencing constraints in G. Therefore, activating the operations based on this schedule will satisfy the precise transition criterion if we can ascertain that the signal $C_a(j)$ from the offset control for anchor a is asserted when at least j cycles have elapsed since the completion of a. From the state transition diagram of the offset control in Figure 8.10, the offset FSM remains in $s0$ until the assertion of $done_a$, whereupon it starts sequencing through the states. This implies that the FSM is in state s_j when j cycles have elapsed after the completion of a. By definition, $C_a(j)$ is asserted if the FSM is in any of the states $s_j, s_{j+1}, \ldots, s_{\sigma_a^{max}}$. Therefore, it is asserted when at least j cycles have elapsed after the assertion of $done_a$.

We now show that the precise restarting criterion is satisfied. The $complete_a$ signal for any anchor $a \in A$ is asserted during the final cycle of its execution. Furthermore, $dsink_a$ is asserted if its completion may result in the completion of the entire graph. With these two results in mind, let us consider the expression for $done_G$ in Equation 8.20 that is composed of offsets from anchors to the sink vertex v_n. We see that if the offset from $a \in A(v_n)$ is positive, then $done_G$ depends on one previous cycle in the offset control $C_a(\sigma_a(v_n) - 1)$. On the other hand, if the offset is zero, then $done_G$ depends on $C_a(0)$. In this case, a is by definition a direct-sink, and hence $done_a$ is asserted during the final cycle of execution. We conclude that $done_G$ is asserted during the final cycle of execution of the entire graph, the control is resetted during the final cycle of execution, and the precise restarting criterion is satisfied.

Since the precise transition and precise restarting criteria are both satisfied for a particular input sequence, and since $stateless_a$ and $dsink_a$ for all anchors are dynamically evaluated, the relative control implementation for G is precise with respect to the schedule $\Omega(G)$. ||

8.3 Summary of Control Generation

Relative control synthesis maps a behavioral hardware description that can be modeled as hierarchical sequencing graphs into an interconnection of interacting finite-state machines. The features of the approach include the support for data-dependent delay operations, concurrency, hierarchy, and detailed timing constraints. In addition, the resulting control implementation is precise by yielding the minimum control delay for a given schedule. This implies no delay penalties are incurred for the arbitrary nesting of procedure calls, conditionals, and loops in the input behavioral description. The final control unit is specified in terms of logic expressions and registers, described in the SLIF format.

Chapter 9

Relative Control Optimization

The control implementation resulting from relative control synthesis can be optimized further to reduce its size. Control optimization can be performed either at the logic level by using a *finite-state machine* model [MSBSV90, SDP90] or at a higher level by using a hardware model described in terms of constraints on the sequencing and timing of the operations [Wol90, DM92]. In the former case, the operations are bound to control states. This implies that the cycle-per-cycle behavior of the control cannot be altered without changing the control specification. Techniques such as sequential logic synthesis and microcode compaction can be used to reduce the cost of the control implementation [GRVD87].

In contrast, in the latter approach, hardware behavior is modeled as a set of sequencing and timing constraints on the operations; the activation of an operation is synchronized to the completion of a set of operations. Since operations are not bound to control states, it is possible to modify the activation time of an operation provided the sequencing dependencies and timing constraints of the original specification are still satisfied. The wider latitude in choosing among a set of possible implementations can lead to a more efficient control implementation in terms of area, which can be further improved by logic synthesis techniques.

In this chapter, we present novel algorithms for control optimization operating at the sequencing graph level. Starting from a constraint graph model of hardware behavior that is derived from sequencing graphs under detailed timing constraints, control optimization modifies the schedule by delaying and serializing operations to reduce the size of the control implementation, generated with the relative control approach. Any modification must guarantee that the original timing constraints remain satisfied. The relative scheduling formulation forms the theoretical base to determine whether a particular schedule modification is valid or not. Because the optimizations are performed at the constraint graph level, the relative control approach of Chapter 8 is directly applicable in generating control from an optimized constraint graph.

The chapter is organized as follows. We describe the criterion that will drive the control optimization in Section 9.1. We then present the concept of *synchronization redundancy* in Section 9.2, and show how they can be used to reduce the control cost. Finally, we present in Section 9.3 a technique called *control synchronization* that introduces redundancies by lengthening and serializing the constraint graph hardware model.

9.1 Control Optimization Criterion

To define the cost function that will drive the control optimization, we briefly outine the mapping from a constraint graph model of hardware behavior to a control implementation. The details are presented in Chapters 6 and 8; we summarize in this section the major results as background for defining the control optimization criterion.

The mapping of a constraint graph to a control implementation involves two tasks: scheduling and control generation. Given a constraint graph, scheduling determines the start times of the operations such that the timing constraints are satisfied. These start times are then used by control generation to derive an FSM specification of the control.

In relative scheduling, the activation of an operation is specified with respect to the completion of a set of *anchors* with data-dependent execution delays. In particular, the *start time* $T(v)$ for a vertex $v \in V$ is defined in terms of the anchor offsets $\Omega = \{\sigma_a(v) \mid a \in A(v)\}$, where $\sigma_a(v)$ represents the time at which v can begin execution after the completion of a, and $A(v)$ represents the anchors in the fan-in of v.

$$T(v) = \max_{a \in A(v)} \{T(a) + \delta(a) + \sigma_a(v)\}$$

$\delta(a)$ denotes the (data-dependent) execution delay of the anchor a. A minimum relative schedule Ω_{min} is one with minimum offsets, which guarantees the minimum overall latency $(T(v_n) - T(v_0))$ for all profiles of execution delays $\{\delta(a), \forall a \in A\}$.

For a particular anchor a, σ_a^{max} represents its *maximal offset* and is given by the equation:

$$\sigma_a^{max} = \max\{\sigma_a(v) \mid a \in A(v),\ \forall v \in V\}$$

This number represents the largest offset for anchor a needed by some vertex v to compute $T(v)$. For example, in Figure 4.7 the maximal offset of anchor b is $\sigma_b^{max} = \max\{\sigma_b^{min}(d),\ \sigma_b^{min}(e)\ ,\ \sigma_b^{min}(t)\} = \max\{3, 5, 5\} = 5$, which is needed for the computation of $T(e)$ and $T(t)$. The maximal offset of an anchor is of interest because it is used to estimate hardware costs as described later.

Given a relative schedule, the control implementation approach of Chapter 8 generates a control circuit to activate operations according to the schedule. The control circuit can be modeled as consisting of two components: an *offset control* for each anchor $a \in A$ and a *synchronization control* for each vertex $v \in V$, as described below:

- The *offset control* for a generates the time offsets with respect to the completion of a. We abstract the offset control as an FSM that is activated by the assertion of $done_a$. The FSM generates a set of signals $C_a(i), 1 \leq i \leq \sigma_a^{max}$, where $C_a(i)$ is asserted when *at least* i cycles have elapsed after the completion of a.

- The *synchronization control* for v synchronizes the activation of v, denoted by $enable_v$, to offsets from the completion of its anchors. Specifically, the enable signal is defined as $enable_v = \prod_{a \in A(v)} C_a(\sigma_a(v))$.

We consider here two possible alternatives for control generation, as shown in Figure 8.9. In the first case, binary counters, which are initially cleared, are used to count the number of cycles since the completion of the anchors; comparators are then used to determine when the operation should be activated. The offset control in this case corresponds to the binary counter and the synchronization control corresponds to the comparators. For the second case, shift registers are used instead of binary counters. This results in more registers in the offset control but reduces the synchronization control since the need for comparators has been eliminated.

In general, many different implementations exist for the control logic. They are obtained by modeling the control as a FSM and applying different state encodings, as illustrated in Figure 8.10. The two implementation alternatives in Figure 8.9 correspond to the two limiting cases of binary and one-hot encodings.

9.1.1 Defining the cost function

Given a specification of control logic in terms of FSMs, we estimate the total **control cost** $COST_{area}$ of the control implementation as:

$$\begin{aligned} COST_{area} &= \sum_{a \in A} COST_{off}(a) &+& \sum_{v \in V} COST_{sync}(v) \\ &= \alpha \cdot \sum_{a \in A} f_{off}(\sigma_a^{max}) &+& \beta \cdot \sum_{v \in V} f_{sync}(|A(v)|) \end{aligned}$$

The first term $COST_{off}(a)$ is the cost due to the offset control for an anchor $a \in A$ and is related to the cost due to the length of the schedule corresponding to a. The cost is a function f_{off} of the maximal offset value σ_a^{max} for a that yields the number of registers implementing the offset FSM for a. The second term $COST_{sync}(v)$ is the synchronization control cost for $v \in V$ and depends on the number of signals that must be synchronized to activate v. The cost is a function f_{sync} of the cardinality $|A(v)|$ of the anchor set of v. The values α and β represent appropriate weight factors related to the actual cost of the logic implementation.

For the two alternate implementation styles shown in Figure 8.9, α is the cost of a register, and β is the cost of a literal. In the counter-based implementation of Figure 8.9(a), the complexity of the offset control is a logarithmic function of the maximal offset values, i.e. $f_{off}(n) = \lceil \log_2(n) \rceil$ represents the number of registers. The complexity of the synchronization control is a linear function of the number of comparators, i.e. $f_{sync}(n) = n$ represents the number of literals in the required comparators. In the shift-register based implementation of Figure 8.9(b), f_{off} is now a linear function of the maximal offset values, i.e. $f_{off}(n) = n$. Since comparisons are no longer needed in the synchronization, $f_{sync}(n)$ represents the number of literals in a n-input AND gate.

In all formulations, the control complexity can be reduced by either minimizing the *maximal offsets* and/or by reducing the size of the *anchor sets*. Our approach to control optimization focuses on these two aspects. Such reductions can be achieved by *modifying* the graph topology and/or the edge weights. These techniques will be described in the following sections.

9.1.2 Control optimization objective

The use of a minimum schedule may, in some cases, result in a higher control cost when compared to a non-minimal schedule (minimal schedules were presented in Chapter 6). The topic of this chapter to find a schedule such that the control cost is minimized while still satisfying all of the constraints.

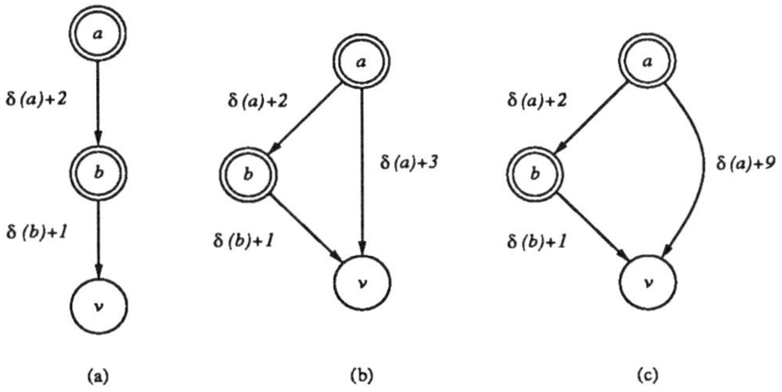

Figure 9.1: Example of redundancy: (a) anchor a is both redundant and irrelevant with respect to vertex v, (b) a is redundant but relevant with respect to v, (c) a is both irredundant and relevant with respect to v.

For a given constraint graph, obtaining different non-minimal schedules can be achieved in one of several ways. We formulate this task as selectively serializing and delaying operations in the original constraint graph model. At each step of the optimization, a new constraint graph G' is obtained that is a slightly delayed version of the original graph G, such that the control cost is reduced, i.e., $COST_{area}(G') \leq COST_{area}(G)$.

The relative scheduling formulation provides the theoretical basis to determine whether the resulting constraint graph is valid. We assume that the input to the control optimization process is a well-posed constraint graph. The result of optimization will maintain this well-posedness property. From Section 6.2.2, Lemma 6.2.3 states that a feasible constraint graph can always be made well-posed as long as there are no cycles with data-dependent weights. Since the original graph is well-posed (and hence feasible), we need only to ensure that no data-dependent length cycle is created in the resulting graph.

9.2 Redundancy in Synchronization

We review now the concept of *anchor redundancy* in a given schedule. Further details, along with theorems and their proofs, are presented in Chapter 6.

It is often the case that some anchors in the anchor set of a vertex v are not needed in the evaluation of its start time, i.e. $T(v)$ is unchanged if offsets from these anchors are not used in its computation. We illustrate the concept of redundancy through the example in Figure 9.1, containing two anchors a

and b and a vertex v. A minimum schedule of v in Figure 9.1(a) and (b) is $\{\sigma_a(v) = 3, \sigma_b(v) = 1\}$, meaning v can execute once 3 cycles have elapsed after the completion of a and 1 cycle after the completion of b. The start times for $T(v)$ and $T(b)$ are given below.

$$\begin{aligned} T(v) &= \max\{T(a) + \delta(a) + 3,\ T(b) + \delta(b) + 1\} \\ T(b) &= \max\{T(a) + \delta(a) + 2\} \end{aligned}$$

Substituting the value of $T(b)$ into $T(v)$, we get:

$$\begin{aligned} T(v) &= \max\{T(a) + \delta(a) + 3,\ T(a) + \delta(a) + 2 + \delta(b) + 1\} \\ &= T(a) + \delta(a) + 2 + \delta(b) + 1 \\ &= T(b) + \delta(b) + 1 \end{aligned}$$

Thus we conclude that the offset $\{\sigma_a(v) = 3\}$ is not necessary to compute $T(v)$, and hence a is redundant with respect to v. In contrast, all edges are necessary in Figure 9.1(c). Removal of any edge would result in a violation of the minimum sequencing constraints for some values of $\delta(a)$ and $\delta(b)$. In this case, we say that anchor a is irredundant with respect to v, which means that offsets from a must be used to compute $T(v)$.

The irredundant anchor set $IR(v)$ of a vertex v is the minimum set of synchronizing points affecting the activation of v. In Figure 9.1(b) the irredundant anchor sets are $IR(v) = b$ and $IR(b) = a$. The proof stating the equivalence of start times computed with and without the redundant anchors is given in Chapter 6.

By using only irredundant anchors in computing the start times, the control cost can be reduced by (1) reducing the size of the anchor sets, translating to lower synchronization costs, and by (2) reducing the maximal offset values, translating to fewer number of states in the corresponding FSM.

The basic idea of the control optimization strategy is to not only remove the redundancies in the original input behavior, but to *introduce* additional redundancies into the original behavior by delaying and serializing operations, with the requirement that the original timing and sequencing constraints remain satisfied.

9.2.1 Making Anchors Redundant

Consider an anchor a that is irredundant with respect to a vertex v. It is sometimes possible to make it redundant either by *lengthening* an existing path or by *serializing* to introduce a new sequencing dependency in the constraint graph. We define a serialization of a graph G to be the creation of a new sequencing

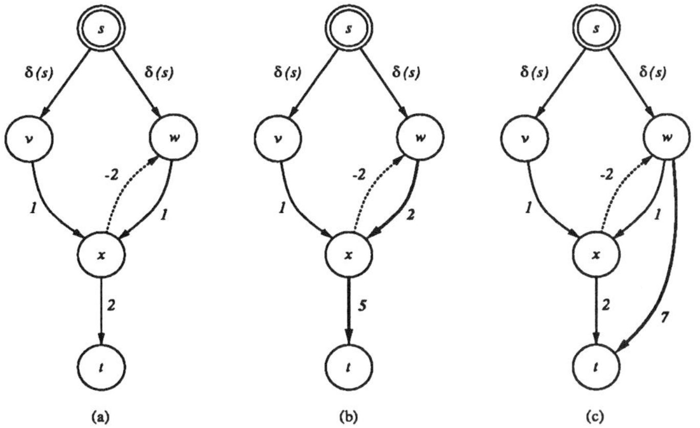

Figure 9.2: Path lengthening: (a) original graph (b) lengthen path $\rho(w, t)$ from 3 to 7 by increasing existing edge weights (c) lengthen by adding edge e_{wt} and increasing its edge weight

dependency that did not exist in the original graph. Lengthening a graph G is defined to be the increase of a minimum timing constraint between two vertices.

In practice, lengthening a constraint between two vertices w and t can be achieved in two different ways, as shown in Figure 9.2. The obvious method, illustrated in Figure 9.2(b), is to increase existing edge weights on edges (shown in bold) of a path $\rho(w, t)$. An alternative method, shown in Figure 9.2(c), consists of the following two steps. First an edge e_{wt} that has an associated edge weight $w_{wt} = |lp(w, t)|$ is added to the constraint graph. This edge weight is then increased by the appropriate amount to achieve the desired lengthening. The second method is used at times to eliminate the unwanted side effect of lengthening other paths unnecessarily. For example, the path $\rho(v, t)$ is lengthened as a side effect in Figure 9.2(b).

We illustrate in Figure 9.3 how an anchor a can be made redundant with respect to vertex v using the serializing and lengthening techniques. The original graph in Figure 9.3(a) is modified by adding a serialization edge e_{ab} from a to b in Figure 9.3(b). The addition of this edge makes the edge e_{sb} redundant. The edge e_{ab} is then lengthened in Figure 9.3(c) to make the edge e_{av} redundant. Since a is now redundant with respect to v, the edge e_{av} can be removed from the graph without changing the constraints on $T(v)$. Because $T(v)$ no longer explicitly depends on the completion of anchor a, the control cost is reduced in

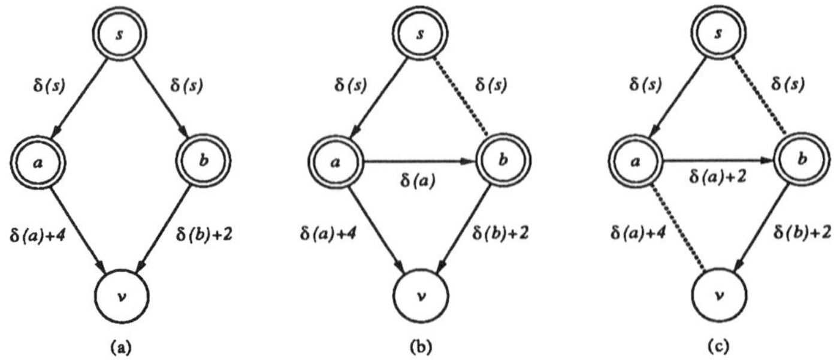

Figure 9.3: Introducing redundancy: (a) original graph (b) serialization edge e_{ab} added (c) edge e_{ab} lengthened to make a redundant with respect to v (redundant edges are dimmed)

two ways. For the graph in (a), the start time is given by:

$$T(v) = \max\{T(a) + \delta(a) + 4, T(b) + \delta(b) + 2\}$$

However, for the graph in (c), the start time is reduced to:

$$T(v) = T(b) + \delta(b) + 2$$

Since this calculation is done dynamically during execution, the synchronization logic for v is reduced. Furthermore, the maximal offset σ_a^{max} is reduced from 4 to 2 which results in a smaller number of registers needed by the offset generation for a.

The delaying of operations via serialization and lengthening must be carried out with care to avoid violating the constraint graph, i.e. making the resulting graph ill-posed. Both lengthening and serializing can be viewed as adding a forward edge e_{av} representing a sequencing constraint from anchor a to vertex v. This edge has a non-negative edge weight $w_{av} = \delta(a) + k$, $k \geq 0$. If a path already exists from a to v containing $\delta(a)$, then the new edge implies lengthening; otherwise, it implies serializing.

Before proceeding, we state the following theorem, which provides the basis for determining the validity of transformations. They relate the concept of feasibility (satisfaction of constraints when $\delta(a) = 0$, $\forall a \in A$) to the concept of well-posedness (satisfaction of constraints for any $\delta(a)$, $\forall a \in A$).

Theorem 9.2.1 *Consider a well-posed constraint graph $G(V, E)$. If a forward edge e_{av} from anchor a to vertex v with weight $w_{av} = \delta(a) + k, k \geq 0$ is added*

from anchor a to vertex v, then the resulting graph $\widetilde{G}$ can be made well-posed if and only if:
(1) $\widetilde{G}$ is feasible, and
(2) $\widetilde{G}$ does not contain unbounded length cycles.

Proof: This follows directly from Theorem 6.2.3. For the sufficient condition, if $\widetilde{G}$ is feasible and does not contain unbounded length cycles, then from Theorem 6.2.3 we know that $G(V, E)$ can be made well-posed. Now for the necessary condition. If $\widetilde{G}$ can be made well-posed, then it is feasible. Furthermore, Theorem 6.2.3 implies that no unbounded length cycles exist in $\widetilde{G}$. ||

This theorem ensures that a constraint graph can be made well-posed provided no positive cycles and no unbounded length cycles are created during control optimization. Furthermore, if it can be shown that the anchor containment property holds, then it follows that the resulting graph *is* well-posed.

9.2.2 Prime versus Non-prime Anchors

If an operation is delayed by *lengthening* existing paths in the constraint graph without introducing new serializations, then the anchor sets of the vertices remain unchanged. The reason is because the transitive sequencing dependencies are not affected by lengthening paths. It is useful to identify the subset of anchors in the anchor set of a vertex which can be made redundant by lengthening.

Definition 9.2.1 *An anchor $a \in A(v)$ of a vertex v is* **prime** *if $a \notin A(q)$, $\forall q \in A(v)$. Otherwise, the anchor is* **non-prime**. *The set of prime anchors of a vertex v is the* **prime anchor set** *of v, denoted by $PA(v)$.*

This means that for a prime anchor $a \in PA(v)$, all paths of forward edges from a to v contain no unbounded delay weights other than the delay of the prime anchor $\delta(a)$. From the definition of anchor redundancy, it follows that a prime anchor is irredundant. For example, in Figure 9.3(a), both a and b are prime anchors of vertex v, because there exists no other anchors on any path from a or b to v. On the other hand, anchor a in Figure 9.3(b) is *not* a prime anchor of v, because there exists a forward path from a to v that contains an unbounded edge weight other than $\delta(a)$, namely $\delta(b)$. In this case it is possible to lengthen the graph to make a redundant with respect to v, as shown in Figure 9.3(c).

Given a vertex v and an irredundant anchor $a \in A(v)$, if a is non-prime, it can be made redundant by lengthening a path from a to v such that no timing constraints are violated. A constraint graph is called **taut** if all non-prime anchors of a vertex are made redundant with respect to the vertex, for all vertices. We state the following lemma.

Lemma 9.2.1 *Consider a vertex $v \in V$ of a well-posed graph, a non-prime anchor $a \in A(v)$, and a prime anchor $b \in A(v)$. The anchor a can always be made redundant with respect to v by lengthening the minimum timing requirements from a to b and from b to v. Furthermore, the resulting constraint graph remains well-posed.*

Proof: We prove by construction. First, let P denote the set of all paths from a to v containing an edge with weight $\delta(a)$. If a longest path $lp(a,v) \in P$ contains an unbounded delay weight other than $\delta(a)$, then a is already redundant. Otherwise, a path $\rho(a,b,v) \in P$ containing $\delta(b)$ will be lengthened to make a redundant. This lengthening will be done in two steps. First the edges e_{ab} and e_{bv} are added with the weights $w_{ab} = \delta(a)$ and $w_{bv} = \delta(b)$. Then these weights are increased so that $|\rho(a,b,v)| = |lp(a,v)|$, thus making a redundant with respect to v.

Assume first that there do not exist paths $\rho(b,a)$ and $\rho(v,b)$. Then the corresponding edge is *unconstrained* and its weight can be made arbitrarily large. This allows for $\rho(a,b,v)$ to be lengthened to $|lp(a,v)|$ and the proof is complete. On the other hand, if both edges are constrained, then their weights can be increased to the values $w_{ab} = \delta(a) + |lp(b,a)|$ and $w_{bv} = \delta(b) + |lp(v,b)|$ so that two zero length cycles are created. These cycles have zero length because forward paths $\rho(a,b)$ and $\rho(b,v)$ exist, which implies that both $|lp(b,a)|$ and $|lp(v,b)|$ are negative.

For the sake of contradiction, assume that this lengthening does not make a redundant. Since a is not redundant, it follows that $l = |\rho(a,b,v)| < |lp(a,v)|$. However, a zero length cycle exists between a and v which implies that there exists a path of length $-l$ from v to a. By combining this path of length $-l$ from v to a along with the path $lp(a,v)$, a positive cycle exists since $l < |lp(a,v)|$. Since no positive length cycles are created, it follows that the original graph contains a positive cycle. By contradiction a must be redundant. Hence, a lengthening does exist. Furthermore, since no positive cycles are created and the anchor sets for all vertices remain the same, it follows that the resulting graph is well-posed. ||

The above lemma implies that any non-prime anchor can be made redundant with respect to its associated vertex. Lengthening can be applied repeatedly until all of the non-prime anchors of a vertex $v \in V$ have been made redundant. We now show that there exists an ordering of the anchors such that a lengthening to make one anchor redundant does not cause a previous anchor in the ordering to become irredundant.

Theorem 9.2.2 *Given a well-posed constraint graph $G(V, E)$, there always exists a lengthening of G, denoted by $\widetilde{G}$, such that $\widetilde{G}$ is taut and well-posed.*

Proof: We prove by construction. The graph is traversed in topological order starting from the source vertex. When an anchor $a \in A$ is traversed it is made redundant to all vertices to which is non-prime and irredundant. We claim that making a redundant in this manner will not cause a previous anchor $c \in A$ in the ordering to become irredundant with respect to some vertex to which it is non-prime.

Suppose anchor a is to be made redundant with respect to a vertex v. According to Lemma 9.2.1, this can be done by increasing unbounded edge weights of the edges e_{ab} and e_{bv} where $b \in A(v)$ is a prime anchor of v. We assume for the sake of contradiction that some previous anchor c in the ordering has inadvertently been made irredundant with respect to some vertex w. That is, the length of a longest path $lp(c, w)$ from c to w has increased. Since c is now irredundant with respect to w, it follows that the path $lp(c, w)$ contains exactly one unbounded delay weight $\delta(c)$. Every path that has been increased must contain at least one of the unbounded weights $\delta(a)$ or $\delta(b)$ because lengthening is accomplished by increasing the weight on an edge from one these anchors. Hence, this implies that $c = a$ or $c = b$. However, this means that c has not yet been traversed. Thus, by contradiction of the assumption, the proof is complete. It follows from Lemma 9.2.1 that the resulting graph is well-posed. ||

We point out that the prime anchor sets are *fixed* for a given graph topology. Furthermore, they remain unchanged when the graph is lengthened. However, it is possible to change the prime anchor sets by serializing the graph, which alters the transitive sequencing dependencies. We use a combination of graph serialization and graph lengthening (as shown in Figure 9.3) to introduce redundancy, which can be removed to optimize the control. This approach is described next.

9.3 Control Optimization Approach

From the previous section, we see that synchronization redundancies can be used to reduce the control cost. Since redundant anchors do not affect the start time of an operation, only the irredundant anchors of an operation represent its *synchronization points*, i.e. they synchronize the execution of an operation with respect to multiple concurrent execution flows.

We now formulate the task of control optimization as *minimizing the control cost* $COST_{area}$ *by modifying the constraint graph*, where the modification is modeled as either graph lengthening, graph serialization, or a combination of both. We consider any modification to the constraint graph to be acceptable as long as the resulting graph remains well-posed. Note that since we never remove any dependencies or constraints in the original graph, the resulting graph still satisfies the original behavior specification.

The two factors in the control cost, synchronization and offset costs, are tightly coupled. The reduction of one may result in the increase of the other. Furthermore, a globally minimum solution does not necessarily imply minimum values in both synchronization and offset control costs. Since simultaneous minimization of both factors may be computationally hard to solve exactly, we use the following heuristic strategy. This strategy is based on the observation that non-prime anchors can always be made redundant, and that a prime anchor of a vertex can be made non-prime by serializing it with respect to another prime anchor.

1. *Minimize the prime anchor sets* – by serializing the anchors.

2. *Resynchronize operations with respect to new synchronization points* – by serializing the operations with respect to the anchors to minimize the offset control cost.

3. *Remove redundancy* – by lengthening the graph to make it taut, which minimizes the synchronization control cost.

Although alternative optimization strategies exist, this approach has the advantage of being able to yield a globally minimum solution for a subclass of constraint graphs. The three steps in this approach are now described in the following sections.

9.3.1 Minimize Prime Anchor Sets

We first serialize the anchors to make as many anchors of a vertex non-prime as possible. Since non-prime anchors can be make redundant by lengthening, this has the effect of reducing the synchronization cost of the final control implementation. This is particularly important for control-dominated designs because

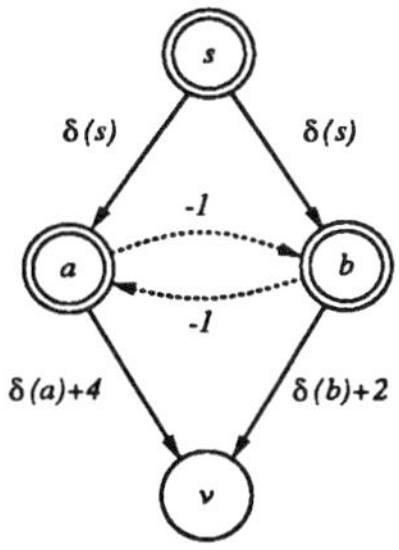

Figure 9.4: Example of anchors that cannot be serialized.

the synchronization cost may dominate the overall control cost in certain implementation styles.

The lower bound on the synchronization control cost corresponds to the case where every vertex, excluding the source, has a single synchronization point, i.e. $\sum_{v \in V} |IR(v)| = |V| - 1$, where $|IR(v)|$ is the number of irredundant anchors of v. In this case, the anchors are *completely* serialized with respect to one another. In the presence of maximum timing constraints, however, it is not always possible to arbitrarily serialize two anchors without violating timing constraints. Specifically, two anchors connected by a cyclic timing relationship cannot be serialized because an unbounded length cycle would be formed. This is illustrated in Figure 9.4 where anchors $a, b \in PA(v)$ cannot be serialized and, therefore, cannot be made non-prime with respect to v. To address this issue, we define an *anchor cluster* as follows.

Definition 9.3.1 *An* **anchor cluster**, *denoted by* λ_i, *is a maximal subset of strongly connected anchors in the constraint graph.*

The set of anchor clusters is denoted by $\{\lambda_0, \lambda_1, \ldots, \lambda_k\}$, where λ_0 is the cluster containing the source vertex. A constraint graph is called **elementary** if all anchor clusters contain a single anchor, i.e. $|\lambda_i| = 1, \forall i$.

Since strong connectivity is an *equivalence relation*, the anchor clusters form a partition over the set of anchors A. Furthermore, the set of anchor clusters form a *partial order*, and it is possible to find a serialization of the anchor clusters such that the clusters are completely ordered. In Figure 9.4, there are two anchor clusters $\lambda_0 = \{s\}$ and $\lambda_1 = \{a, b\}$, and they are completely ordered. In the case where all clusters contain a single anchor, an ordering results in a chain of anchors from source to sink. More formally, we define a *cluster ordering* as follows.

Definition 9.3.2 *A* **cluster ordering** *of a constraint graph G is a complete serialization of the anchor clusters of G, such that for every pair of clusters λ_i and λ_j, every anchor $a \in \lambda_i$ is serialized with respect to every anchor $b \in \lambda_j$. The graph G with a cluster ordering is called an* **ordered** *graph.*

Since only the irredundant anchors are necessary in the start time computation of a vertex, serializing the clusters can reduce the synchronization requirement by decreasing the number of irredundant anchors for a vertex. In order to define a lower bound on the synchronization cost, we present the following theorem. It applies to those graphs that are ordered.

Theorem 9.3.1 *For a well-posed, ordered, elementary constraint graph, the sum of the cardinality of the prime anchor sets for all vertices is equal to $|V| - 1$.*

Proof: $|PA(v)| = 1$, $\forall v \in V$ except $PA(v_0) = 0$. This implies that $\sum_{v \in V} |PA(v)| = |V| - 1$. $||$

Since we can make a graph taut, the theorem above implies that it is possible to achieve the lower bound in synchronization costs for elementary graphs. We note that imposing a cluster ordering in a graph will not affect the property of well-posedness. The reason is that, by definition, anchor clusters are not connected by any cycle in the constraint graph. Therefore, no cycles can be formed by serializing between anchors in different clusters.

The search for a cluster ordering that is compatible with the original partial order can be carried out using branch-and-bound techniques. Alternatively, heuristic algorithms can be applied to limit the search for a good solution. We use a heuristic that orders the anchor clusters by ranking them with respect to increasing lengths of the longest path from the source vertex. This ranking is then used to convert the original partial order to a complete order by serializing the clusters. Cluster serialization is done by adding sequencing dependencies from the anchors in one cluster to those anchors in the following cluster. As noted above, this serialization step does not create any cycles. Furthermore the anchor containment property is preserved because edges are added only between anchors. For example, in Figure 9.8(a) the longest paths from the source anchor a to the three anchors b, c, and d are computed. This results in the cluster ordering indicated by the shaded arrows.

9.3.2 Partition to Resynchronize Operations

Informally a cluster ordering can be viewed as a *chain* of clusters; each link in the chain represents a set of synchronization points along with a set of operations that depend on these points. Given a cluster ordering, we *partition* the operations and assign them to the links of this chain in order to minimize the offset control cost. The process of assigning operations to links involves serializing the operations with respect to the anchor clusters, and it is called **resynchronization** since the activation of operations will depend on a possibly new set of synchronization points. We formalize this idea with the following definition.

Definition 9.3.3 *Given a well-posed, ordered constraint graph with two consecutive anchor clusters λ_i and λ_{i+1}, consider all paths from the anchors in λ_i to the anchors in λ_{i+1} such that each path has a single unbounded edge weight corresponding to an anchor in λ_i. The subgraph induced by the vertices on these paths is defined to be a* **cluster link** *$\mathcal{L}_i$ corresponding to λ_i. A* **segmented** *graph is a graph for which all vertices belong to cluster links.*

A vertex v belonging to a cluster link $\mathcal{L}_i$ satisfies the condition that it is a successor to at least one anchor in λ_i, and therefore its prime anchor set is equal to a subset of the anchors in λ_i, i.e. $PA(v) \subseteq \lambda_i$. A segmented graph is a special form of ordered graph where all operations belong to links.

In general, not all vertices of an ordered graph belong to cluster links. For example, a vertex may not fall between consecutive clusters, but rather lie on a path between the anchors of non-consecutive clusters. Consider the ordered, elementary graph in Figure 9.8. In (a), only v_2 and v_5 belong to a link, whereas in (b), all vertices belong to links. The objective is to partition the vertices by assigning them to the links in such a way as to minimize the overall offset control cost.

Due to the presence of maximum timing constraints, the vertices cannot be assigned arbitrarily. We define a **constrained operation-set** (c-opset) as a maximal subset of strongly connected vertices, so called because the execution of these vertices are constrained with respect to each other. For example, in Figure 9.8(a), the vertices $\{v_3, v_4\}$ form a c-opset. A trivial c-opset is one containing a single vertex. We state without proof that all elements of a c-opset must be assigned to the *same* link in the chain if the resulting graph is to be well-posed. If two operations in a c-opset belonged to different links, a positive unbounded cycle would exist.

We now present a heuristic algorithm, called *ResyncGraph* shown in Figure 9.5, that assigns c-opsets to the links of the cluster chain. The anchor clusters are traversed in topological order. For each cluster λ_i, all of the vertices v are examined that have a prime anchor in the cluster (i.e. $a \in PA(v)$ and $a \in \lambda_i$). If there exists a path from v to any anchor in the following cluster λ_{i+1}, then v

```
ResyncGraph(G)
  {
    /* iterate over clusters */
    foreach ( cluster λ_i )
      foreach ( vertex v such that PA(v) ⊆ λ_i )
        if ( a path exists from v to λ_{i+1} )
          /* assign v to cluster λ_i */
          foreach ( anchor a ∈ λ_{i+1} )
            Add edge e_va
            if ( |lp(v,a)| > 0 )
              w_va = -|lp(v,a)|
            else
              w_va = -0
        else
          W = {w | a path ρ(v,w) of backward edges exists}
          foreach ( anchor a ∈ λ_{i+1} )
            /* defer assignment of v to cluster λ_{i+1} */
            Add edge e_av with weight w_av = δ(a)
            /* ensure that anchor containment is preserved */
            foreach ( w ∈ W )
              Add edge e_aw with weight w_aw = δ(a)
  }
```

Figure 9.5: *ResyncGraph* Algorithm

must be assigned to the cluster λ_i; otherwise, an unbounded cycle would result. This assignment is done by adding a backward edge from v to each anchor in the cluster λ_{i+1}. Note that a forward path from λ_i to v already exists because the prime anchors $PA(v)$ are contained in λ_i. Appropriate weights are given to the added backward edges so that no positive cycles are created. The reason for adding these backward edges is to delay the cluster λ_{i+1} as much as possible without increasing the maximal offsets for the anchors in the previous cluster λ_i. Consider the graph in Figure 9.6(a), which has two clusters $\lambda_1 = \{a\}$ and $\lambda_2 = \{b\}$. Vertex v must be assigned to cluster λ_1 because there exists a path from v to b in (a). Therefore, the backward edge e_{vb} is added, which has the effect of increasing the edge weight w_{ab} from $(\delta(a) + 0)$ to $(\delta(a) + 4)$. The maximal offset for a does not change and remains at 4.

If there exists no paths from v to λ_{i+1}, then serialization edges are added from the anchors in λ_{i+1} to v. This *resynchronizes* v to a different anchor cluster by changing its prime anchor set so that $PA(v) \subseteq \lambda_{i+1}$. This has the effect of

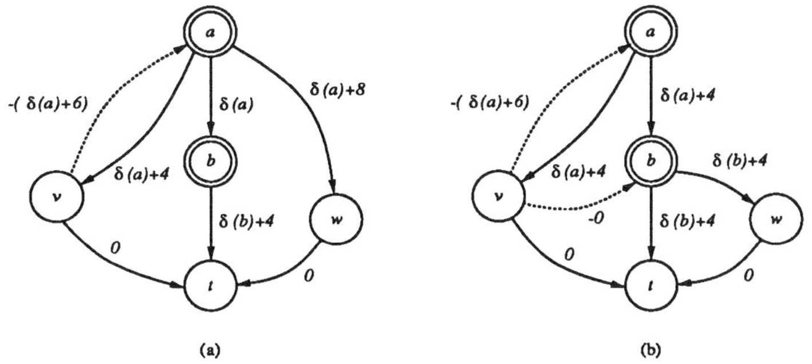

Figure 9.6: Example of resynchronization of operations.

amortizing the offsets needed for $T(v)$ over consecutive anchor clusters. For example, in Figure 9.6(a) no path exists from w to a so the edge e_{bw} is added. This edge is later lengthened to make e_{aw} redundant. The combination of adding backward edges and resynchronizing operations has reduced the maximal offset σ_a^{max} from 8 to 4. This in turn reduces the register requirement needed by the offset generation for anchor a. Neither of these steps by itself would have achieved such a result. In both of these cases edges are added, but the type and weight of an added edge is such that no bounded or unbounded positive cycles are created. Furthermore, forward edges are added in a way that preserves the anchor containment property.

9.3.3 Making the Graph Taut

The final step in the optimization strategy is to remove redundancies by making the constraint graph *taut*. This is done by making the non-prime anchors of a vertex redundant by lengthening the graph appropriately. However, it is possible to increase the maximal offset values as a result of lengthening. It was shown in Section 9.1.1 that such an increase has a direct effect on the hardware control cost. Therefore, an important criterion in lengthening is to minimize the *amount* of increase. The algorithm for lengthening, called *LengthenTaut*, is shown in Figure 9.7.

The algorithm for lengthening visits each anchor a in the graph and lengthens the paths to vertices v, where a is a non-prime irredundant anchor of v, in order to make a redundant. For a vertex v and a non-prime anchor $a \in A(v)$, a forward path $\rho(b_0, \cdots, b_n)$ is found where $b_0 = a$, $b_n = v$, and $b_i \in A(v)$ for $0 \leq i < n$. Furthermore the path contains the unbounded edge weights

```
LengthenTaut(G)
  {
      foreach ( anchor a ∈ A in topological order )
          foreach ( vertex v where a ∈ IR(v) )
              /* find forward path from a to v */
              find ρ(b_0, ···, b_n) : b_0 = a, b_n = v, b_i ∈ A(v), 0 ≤ i < n
              increase = |lp(a, v)| − |ρ(b_0, ···, b_n)|
              for (i = 0; i < n; i++)
                  if ( increase > 0 )
                      c = b_i, d = b_{i+1}
                      if ( e_cd does not exist )
                          Create new edge e_cd with weight w_cd = δ(c)
                      /* lengthen e_cd as much as possible */
                      slack = |lp(d, c)|
                      if ( slack > increase )
                          added = increase
                      else
                          added = slack
                      w_cd = w_cd + added
                      increase = increase − added
  }
```

Figure 9.7: *LengthenTaut* Algorithm

$\delta(b_i)$ for $0 \leq i < n$. This path is lengthened by traversing the anchors of the path starting with a. Edges are added between consecutive anchors, and their weights are increased as much as possible until either a maximum constraint limit is reached, or until a is made redundant with respect to v. Lengthening in this way does not introduce any bounded or unbounded positive cycles. Since lengthening does not introduce a forward path between two vertices where such a path did not previously exist, it follows that the anchor containment property is preserved by this operation.

By Theorem 9.2.2 the algorithm is guaranteed to make a constraint graph taut. In addition, the algorithm also guarantees for *elementary* constraint graphs, the control offset cost is always reduced, or in the worst case remains the same. In order to determine the effect of these steps on the control cost, we state the following lemma and theorem.

Lemma 9.3.1 *Consider a well-posed constraint graph $G(V, E)$ and the set of maximal offsets $S = \{\sigma_a^{max} \mid a \in A\}$. An edge e_{bc} added between two anchors*

b, c with an unbounded edge weight $\delta(b)$ can modify only the maximal offset $\sigma_b^{max} \in S$ provided the resultant graph remains well-posed.

Proof: Only graphs that remain well-posed are considered here because the notion of a maximal offset is not well defined for ill-posed graphs. The maximal offset σ_b^{max} of an anchor b can be computed by finding the longest path from b to all vertices v where $b \in IR(v)$. For the sake of contradiction, assume there exists an anchor $d \neq b$ such that σ_d^{max} has increased as a result of adding e_{bc}. This implies that there exists $c \in A$, $w \in V$, and a path $\rho(d, w)$ (containing a single unbounded delay $\delta(d)$) that has increased in length. Since only the edge e_{bc} has been added, any path that has changed length must contain e_{bc}. Hence the increased path $\rho(d, w)$ contains an unbounded delay weight other than $\delta(d)$, namely $\delta(b)$. By contradiction the proof is complete. Only graphs that remain well-posed are considered here because the notion of a maximal offset is not well defined for ill-posed graphs. ||

Theorem 9.3.2 *Given a well-posed, elementary constraint graph $G(V, E)$, G can be made taut without increasing the maximal offset values of the anchors of G.*

Proof: Consider a vertex $v \in V$ and a non-prime anchor $a \in A(v)$ that is irredundant. Since a is non-prime, there exists a forward path $\rho(a, b)$ to some anchor $b \in PA(v)$ containing the unbounded delay weight $\delta(a)$. By definition of irredundancy, a longest path $lp(a, v)$ from a to v contains only the one unbounded delay weight $\delta(a)$. It follows that $|lp(a, v)| > |\rho(a, b)| + |\rho(b, v)|$. By definition, $|lp(a, v)| \leq \sigma_a^{max}$. An edge e_{ab} can be added with weight $w_{ab} = \delta(a) + |lp(a, v)| - |\rho(b, v)| \leq \sigma_a^{max}$. Such an edge can always be added without making the resulting graph ill-posed. This is assured because a positive cycle could result only if a and b were already strongly connected. However, such a cycle cannot exist because the graph is elementary. Therefore, a has been made redundant without increasing its maximal offset. In fact, its maximal offset may actually have been reduced. According to the Lemma 9.3.1 we know that no other maximal offsets can possibly change, hence the proof is complete. ||

According to the above theorem, given a well-posed, elementary constraint graph G, procedure *LengthenTaut* can make G taut without increasing the maximal offset values of the anchors of G.

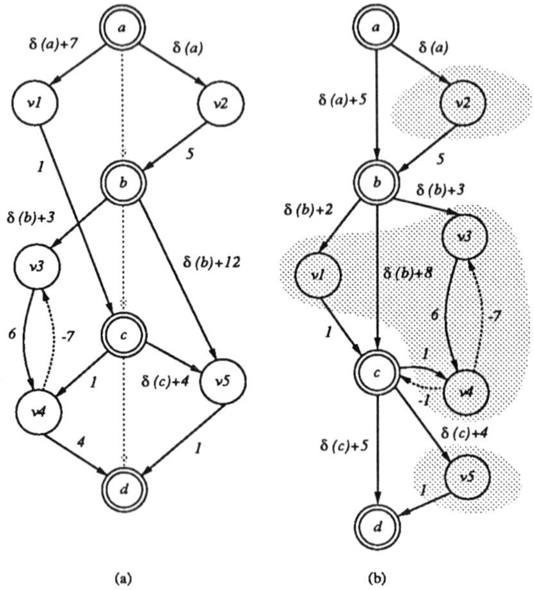

Figure 9.8: Example of (a) an un-segmented constraint graph, where v_2 is the only c-opset belonging to a link, (b) a segmented constraint graph. The shaded arcs in (a) represent the cluster ordering that is derived based on ranking of longest path lengths from the source.

9.3.4 Example

We illustrate the application of our strategy in Figure 9.8. The graph contains four anchors $\{a, b, c, d\}$, and five vertices $\{v_1, v_2, v_3, v_4, v_5\}$, where v_3 and v_4 form a c-opset. The graph is elementary since each anchor cluster contains a single element. The first step is to find the longest path from the source vertex a to all other vertices in the graph. Based on increasing length of these longest paths, a cluster ordering is imposed. This results in the cluster ordering (a, b, c, d) corresponding to the longest path lengths 0, 5, 8, 18 as seen in Figure 9.8(a). This ordering is realized by adding serialization edges e_{ab}, e_{bc}, and e_{cd} with the respective edge weights of $\delta(a)$, $\delta(b)$, and $\delta(c)$.

The next step is to resynchronize the operations. The vertices are grouped into c-opsets, which are then assigned to cluster links. In this example there is only one non-trivial c-opset $\{v_3, v_4\}$. Resynchronization results in the c-opset assignments of $\{v_1\}$ to $\mathcal{L}_b$, $\{v_2\}$ to $\mathcal{L}_a$, $\{v_3, v_4\}$ to $\mathcal{L}_b$, and $\{v_5\}$ to $\mathcal{L}_c$.

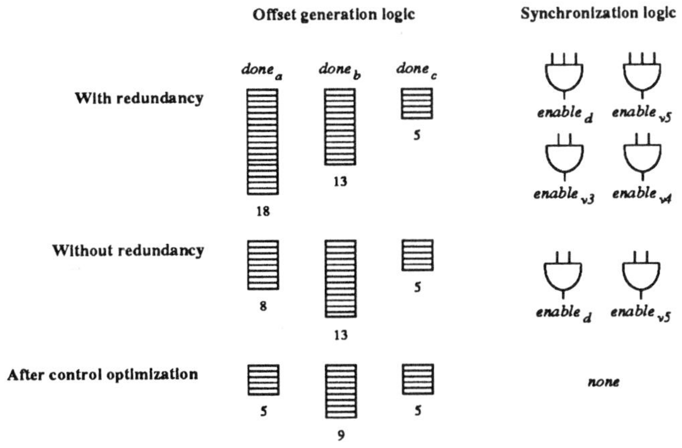

Figure 9.9: Illustration of control costs.

Assignment of a vertex v to a link $\mathcal{L}_i$ is achieved by adding a forward edge from λ_i to v and a backward edge from v to λ_i. For example, v_4 was assigned to its link by adding the edges e_{b,v_4} and $e_{v_4,c}$ with edge weights $w_{b,v_4} = 0$ and $w_{v_4,c} = -1$.

Finally, redundancy is removed from the graph by lengthening it to make it taut. This results in the increased values of the edge weights $w_{ab} = \delta(a) + 5$, $w_{b,v_1} = \delta(b) + 2$, $w_{b,c} = \delta(b) + 8$, and $w_{c,d} = \delta(c) + 5$. Comparing the original graph with redundancies removed to the optimized graph, we see that the offset cost is reduced from $\sigma_a^{max} + \sigma_b^{max} + \sigma_c^{max} = 8 + 13 + 5 = 26$ to $5 + 9 + 5 = 19$. The synchronization cost is also reduced in the optimized graph, were all vertices depend only on a single synchronization point. So in this example, both components of the control cost have decreased. The graphic in Figure 9.9 represents both the offset generation and synchronization costs for an implementation based on shift registers as described in Section 9.1.1. The number of registers needed for offset generation and the synchronization logic are shown in the figure. The first row of numbers corresponds to the original graph. The costs in the second row have been reduced by removing the original redundancies. Finally, applying control optimization results in the final row of numbers.

9.4 Analysis of Control Resynchronization

In general, the interaction between anchor clusters and c-opsets complicates the analysis of the problem formulation and its solution space. For example, it is not guaranteed that a globally minimum cost graph is always ordered or segmented. However, for the case of an *elementary* and *ordered* constraint graph, we can show that it is always possible to find a minimum control cost $COST_{area}$ solution using our formulation. Specifically, we now show that of all the possible minimum solutions for an elementary, ordered graph, at least one of them is segmented. Thus, by searching all possible ways to segment the graph, it is guaranteed that a minimum solution will be found.

Formally, we say a graph $\widetilde{G}$ *implements* another graph G if it is derived from G by serializing and lengthening, and we state the following key theorem.

Theorem 9.4.1 *Given a well-posed, ordered, elementary graph G with control cost* $COST^G$. *A well-posed* **segmented** *graph* $\widetilde{G}$ *can always be found such that* $COST^{\widetilde{G}} \leq COST^G$.

Proof: Consider a vertex $v \in V, v \neq v_0$, of the graph G. By Theorem 9.3.2 we know we can lengthen the graph so that all non-prime anchors of v are made redundant. Furthermore, since G is elementary and ordered we know that $|PA(v)| = 1$. Let $\lambda_0, \lambda_1, \ldots, \lambda_k$ denote the anchor clusters that have been ordered. Since G is elementary, each cluster contains a single anchor. Consider two consecutive clusters $\lambda_i = \{a\}$ and $\lambda_{i+1} = \{b\}$ such that $a \in A(v)$. The length of a longest path $lp(b, v)$ (if one exists) from b to v is denoted by $l_{vb} = |lp(b, v)|$.

In order to assign v to the link $\mathcal{L}_i$, a maximal timing constraint e_{vb} is added. If a path of positive length from b to v exists, then $w_{vb} = -|lp(b, v)|$, otherwise $w_{vb} = -0$. The value of the weight ensures that no positive cycle is formed. Note that l_{vb} is bounded, because if l_{bv} is unbounded, then a cannot be a prime anchor of v, which contradicts our assumption. Since no forward edges are added, the anchor sets for all vertices remain unchanged. Since G is well-posed, no anchor sets have been modified, and no positive cycles have been introduced, $\widetilde{G}$ is well-posed.

We know by definition that $|lp(a, v)|, |lp(a, b)| \leq \sigma_a^{max}$. Adding e_{vb} creates a new path from a to b with length $|lp(a, v)| + w_{vb} \leq |lp(a, v)| \leq \sigma_a^{max}$; thus σ_a^{max} remains unchanged. We now show that the maximal offsets of the other anchors do not increase as a result of the added edge. Consider an anchor d where $a \in A(d)$, i.e. d follows a in the ordering. For the sake of contradiction, assume that σ_d^{max} has increased. This implies a path $\rho(d, v)$ exists from d to v containing the unbounded edge weight $\delta(d)$. Since $a \in A(d)$, it follows that a is not a prime anchor of v. This contradicts the fact the v has been assigned to cluster $\lambda_i = \{a\}$. By contradiction, σ_d^{max} cannot not increase.

Likewise, consider an anchor c where $c \in A(a)$, i.e. c precedes a in the ordering. For the sake of contradiction, assume that σ_c^{max} has increased. This implies that a vertex w exists such that there is a path $\rho(c, w)$ from c to w with exactly one unbounded weight $\delta(c)$. Furthermore, $\rho(c, w)$ must contain the newly added edge e_{vb}. Since paths from c to a and from a to v exist, it follows that a path $\rho(c, a, v, w)$ exists. This path has at least two unbounded weights $\delta(c)$ and $\delta(a)$. This implies that c is not a prime anchor of w, and by contradiction, σ_c^{max} cannot increase.

From the above arguments all maximal offsets remain unchanged, i.e. $COST_{offset}^{\widetilde{G}} = COST_{offset}^{G}$. Since $|PA(v)| = 1$ for all v except for the source vertex, making the graph taut reduces the synchronization cost for each vertex to a minimum, i.e. $COST_{sync}^{\widetilde{G}} = |V| - 1$. Thus $COST_{sync}^{\widetilde{G}} \leq COST_{sync}^{G}$, and $COST^{\widetilde{G}} \leq COST^{G}$. ||

9.5 Summary of Control Resynchronization

We have presented a control optimization strategy based on resynchronization of operations. Using a constraint graph model that supports concurrency, unbounded delay operations, and detailed timing constraints, we showed how the graph can be mapped to a control implementation that consists of synchronization and offset control components. Previous worked focused on reducing the latency of the implementation. Since the minimal latency solution is not always needed, we focused on reducing the control area while at the same time satisfying the timing constraints that can include upper bounds on latency. The result of this approach is an implementation with potentially smaller area while still "fast enough" to meet the specification.

The total control cost is reduced by introducing *synchronization redundancy* in the graph, where any modification to the graph is considered to be acceptable provided the original timing constraints are not violated. We formulate control optimization as a three step process. First selected operations are serialized by adding new sequencing constraints. Operations are then partitioned among the synchronization points (operations with unknown delay). Finally, certain minimum timing constraints are increased in order to remove redundancy and thereby reducing the control cost. Heuristic algorithms were presented based on greedy ranking to improve the computational efficiency at the possible expense of quality.

Chapter 10

System Implementation

This chapter presents an overview of implementation and organization of the *Hercules* and *Hebe* programs [KM90b], which incorporate the formulations and algorithms described in this book. They form the front end to the *Stanford Olympus* Synthesis System for digital design [MKMT90] and are interfaced to the logic synthesis, simulation and technology mapping tools in this vertically integrated synthesis system[1]. Hercules and Hebe have been implemented in approximately 140,000 lines of C code. They are targeted towards the synthesis of synchronous, non-pipelined, digital ASIC designs.

The input to Hercules and Hebe consists of two sets of specifications, a description of the *functionality* and a set of *design constraints*. The functionality is described in the HardwareC hardware description language. The design constraints describe the timing and resource requirements on the resulting synthesized design. They are specified either as part of the HardwareC input description, or interactively by the user. Hercules performs behavioral optimizations on the HardwareC models, producing as output a representation of the optimized behavior in the Sequencing Intermediate Form (SIF). Hebe takes as input the SIF models, performs synthesis tasks such as allocation, scheduling and binding, and produces a synchronous logic-level implementation described in the Structural/Logic Intermediate Form (SLIF). These implementations are called *design points* for the SIF model because they represent different points in the *design space* of tradeoffs between area and performance.

[1] Hercules and Hebe, along with the rest of the Olympus Synthesis System, are publicly available. Information on software distribution can be obtained by sending electronic mail to olympuschronos.stanford.edu from any computer connected on the Internet.

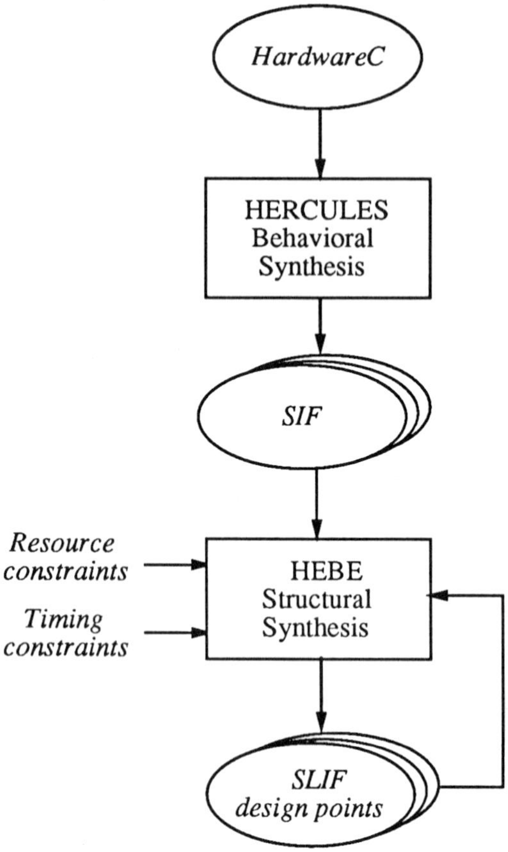

Figure 10.1: Synthesis flow in Hercules and Hebe.

A block diagram of the synthesis flow in Hercules and Hebe is shown in Figure 10.1. There is a one-to-many relationship between a HardwareC model and its SIF models, and between a SIF model and its SLIF design points. We now describe the details of each program. Section 10.1 describes the synthesis flow and organization of Hercules. Section 10.2 describes the synthesis flow and organization of Hebe.

10.1 Overview of Hercules

The *Hercules* system provides an integrated environment for the behavioral synthesis and optimization of hardware behavior. It performs both user-driven and automatic behavioral transformations, and produces as output one or more sequencing graph abstractions of the optimized behavior, described in SIF. The flow of operation in Hercules is described below.

- *Read in HardwareC models.* Several *models* can be read in during a single session. A model corresponds to one of the four design abstractions in HardwareC: *block*, *process*, *procedure*, and *function*.

- *In-line expand model calls.* A model may consist of a hierarchy of calls to other models in the system. For example, an 8-bit adder may call two 4-bit adders cascaded in series. The designer can optionally and selectively replace a model call by its description. By flattening the calling hierarchy, optimization can be performed across the model boundaries.

- *Map Operators to library models.* The description of a model may contain operators such as "+" and "-" that can be synthesized in a variety of implementation styles. For example, an addition can be implemented using either a ripple-carry adder or a carry-lookahead adder. The designer can selectively replace operators by calls to specific models in a particular library. Any operator that is not mapped to a library model will be implemented by default as combinational logic, with the exception of multiply ("*") and divide ("/"), which must always be mapped.

- *Perform automatic behavioral optimizations.* In-line expansion and operator to library mapping are transformations that are guided by the designer. Upon completion of these optional user-driven transformations, a suite of automatic behavioral transformations is performed to optimize the behavior. Note that these optimizations cannot be applied to a *block* model because of its declarative semantics (the list of optimizations is described in Section 3.3). Optimizations include compiler-like optimizations such as dead-code elimination and variable unfolding.

- *Write to SIF output.* Once behavioral optimizations have been performed, the designer can produce a sequencing graph representation of the optimized behavior in SIF.

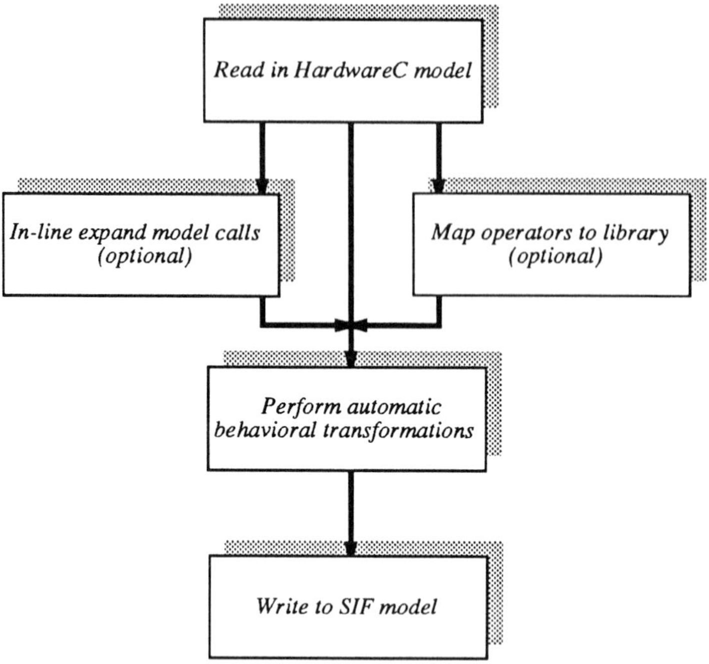

Figure 10.2: Flow of operation in Hercules.

Figure 10.2 illustrates the synthesis flow in Hercules. The organization of models is described in Section 10.1.1. Section 10.1.2 summarizes the command organization in Hercules.

10.1.1 Data organization in Hercules

A HardwareC model that is read by Hercules is first translated into an *abstract syntax tree* representation called the Behavioral Intermediate Format (BIF). The BIF is the underlying representation for all behavioral transformations performed in Hercules. The data models in the system are organized into three levels:

- *Models:* A hardware model in HardwareC is a *prototype* from which *instances* of the model can be created and synthesized. These models are categorized according to their types: block, process, procedure, and function models.

- *Instances:* A model is classified either as a *template* or as a *non-template*. A template takes one or more integer parameters that are used in defining the body of the model. Accordingly, a template model can have one or more *instances*, where an instance corresponds to an unique mapping of values to the integer parameters. For example, the dimensions of variables in the description can be specified as an integer parameter, i.e. an adder template can be defined for operands of any size. In contrast, a non-template model has a default instance corresponding to the description itself. Each model has a *current instance*. The distinction between the description of a model and its instances allows clean separation between the definition and use of a given behavior.

- *Instance Entries:* To allow a designer the flexibility of experimenting with different synthesis options, each instance of a model can have one or more copies, called *instance entries*. The instance entries are uniquely named. Synthesis transformations in Hercules are applied to a specific instance entry. Each instance of a model has a *current entry*.

For each instance entry, Hercules can generates a corresponding SIF model. Therefore, an instance can be mapped to one or more SIF models, with each SIF model corresponding to a particular configuration of user-driven behavioral transformations that was performed on the model. For example, consider a procedure *Main* that calls a function *Func*. If the calling hierarchy is unchanged, then a default SIF model is created. If the call to *Func* is in-line expanded, then the resulting SIF model would be different.

10.1.2 Command organization in Hercules

Hercules is an interactive program with a flexible environment modeled after the Unix^{TM} shell. It provides *command aliasing* and *history* capabilities, as well as input and output file redirection. A consistent user interface is supported by both Hercules and Hebe. The program can be executed either interactively or in batch model through the use of a command script file. Table 10.1 lists the commands that are supported in Hercules, where the commands are divided into the following categories: *system commands*, *model management commands*, *print commands*, and *synthesis commands*.

There are three levels of user complexity in the system: the *novice user*, the *advanced user*, and the *tool developer*. Each level supports a set of commands and the corresponding help facility including those of lower levels. Unsupported

System commands	
alias	alias command string to identifier
cd	change current working directory
current	set current model
echo	echo string to terminal
help	print on-line help
history	print history of previous commands
info	print brief tutorial on system
prompt	change system prompt
quit	exit program
set	set default environment variables
Model management commands	
create	create new instance of model
printmodel	list models in system
readhc	read HardwareC file
reset	delete models from system
restore	undo all optimizations on given model
select	select current instance of model
Print commands	
hierarchy	print calling hierarchy for model
printbif	print BIF format of model
printsif	print SIF format of model
template	print BIF for template model
Synthesis commands	
mapop	maps operator to library template model
bindop	replace operator to calls based on mapping
expand	in-line expand model calls
bsyn	perform automatic behavioral transformations
traverse	traverse BIF hierarchy

Table 10.1: Command summary for Hercules Version 3.0.

commands below a given level are made invisible to the user. This feature assists in getting users acquainted with the tool without overwhelming them with the large number of available options. As designers gain experience in using the tools, the complexity level can be updated accordingly.

10.2 Overview of Hebe

The *Hebe* program takes as input one or more SIF models. Optionally, the designer can specify the following constraints to guide the results of synthesis: *detailed timing constraints* in terms of minimum and maximum bounds on the activation of operations, *resource constraints* in terms of a partial binding of operations to resources and maximum resource usage limits, and *system cycle time* to define the cycle time of the final synchronous hardware implementation. Hebe performs resource sharing to find one or more logic-level implementations of the SIF model such that the imposed constraints are satisfied. The flow of commands in Hebe is described below.

- *Read in SIF models.* Several *SIF models* can be read in during a single session. A SIF model corresponds to either a SIF *structure* model that describes a structural interconnection of models, or a SIF *sequencing* model that describes an imperative semantic algorithm in the sequencing graph model.

- *Manually apply constraints.* The designer can optionally apply the following constraints to a sequencing model G_s: (1) *logic constraints*, by extracting delay and area estimates for the logic expression blocks in G_s, (2) *resource constraints*, by allocating and binding resource instances to calls in G_s, and (3) *timing constraints*, by specifying minimum, maximum, and delay bounds on the operations of G_s.

- *Explore design space.* The designer explores the design space by allocating resources and selecting one or more *resource bindings* as candidates to be synthesized. The selection is guided by both exact and estimated costs on the area and delay of the final hardware. For each candidate resource binding, conflict resolution and relative scheduling are performed to ensure the required timing constraints are satisfied. A *design point* is created and written to the file directory only if timing constraints are satisfied.

- *Synthesize design points.* The design points are synthesized to logic-level implementations by performing control generation and data-path optimiza-

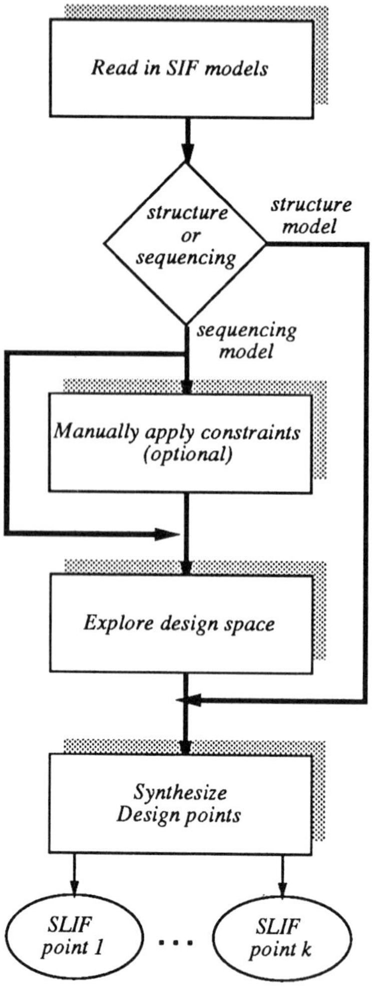

Figure 10.3: Flow of operation in Hebe.

System environment variables	
current	Current SIF model
debug	Enables debugging trace mode
complexity	Complexity level of session
history	# of past commands to remember
verbose	Verbose output message mode
Synthesis environment variables	
cycletime	Cycle time of design
logicscript	Script file for logic synthesis
library	Selected library (e.g. Actel)
maxbinding	Max # resource bindings to generate
maxordering	Max # ordering for conflict resolution
heuristic	Enables heuristic synthesis mode
defalloc	Default allocation (minarea, conflict-free, mindelay)

Table 10.2: Summary of environment variables in Hebe.

tions such as register folding. Each design point can be minimized using logic synthesis techniques and mapped to cells of a given library. The design points can also be simulated to ensure the correctness of the hardware with respect to the original functionality and timing constraints.

Figure 10.3 illustrates the synthesis flow in Hebe. No hardware sharing is permitted for SIF structure models. In this case, the task of Hebe is to select the specific implementation of called models to generate the final logic-level design.

The organization of Hebe is described in Section 10.2.1. The design subsystems are described in Section 10.2.2.

10.2.1 Command organization in Hebe

Hebe provides an interactive and flexible environment for specifying design constraints and exploring the design space. The user interface is similar to Hercules in that it supports command aliasing, file redirection, history capabilities, and different levels of user complexity.

In addition, Hebe also supports an extensive list of *environment variables* that define the default settings and options for the synthesis system. For example, there is an environment variable that sets the cycle time of the design. These variables allow the designer to customize his or her synthesis strategy, which results in a simpler and more usable synthesis tool. The list of environment variables is shown in Table 10.2. Note that the *system environment variables* in the table are also supported by Hercules. We now describe the *synthesis environment variables* below.

- *cycletime:* The designer can set the clock cycle time for a design in absolute time units, usually nanoseconds. If the cycle time is not specified, then by default it is extracted from logic synthesis as the critical path delay in the final logic-level implementation.

- *logicscript:* Logic synthesis, especially technology mapping, is an integral part of the synthesis flow in Hebe. Logic synthesis provides minimization and extraction of delay and area information to guide Hebe's exploration of the design space. The designer can optionally specify a command script file that is executed by the logic synthesis tools. This capability extends the designer's control over the commands and options of the logic-level synthesis tools.

- *library:* The designer can optionally specify a particular library, e.g. Actel or LSI Logic LMA-9K cells, that will be used to evaluate the area and delay characteristic for a given design. If the library is not specified, then the number of logic literals is used as the area estimate and the number of levels of logic is used as the delay estimate.

- *maxbinding:* The designer can limit the maximum number of resource bindings that will be generated in constructing a design space. If this limit is exceeded, then the designer will be queried on whether to continue exhaustively generating the design space, or instead heuristically generate a portion of the design space.

- *maxordering:* The designer can also set the maximum number of valid orderings that conflict resolution will explore. If the maximum ordering is 1, then the first valid ordering is used. Otherwise, the best valid ordering among these orderings in terms of overall latency is selected. If no maximum ordering is specified, the system will exhaustively try all ordering possibilities and select the best one.

- *heuristic:* The heuristic mode selects between the exhaustive and heuristic design space generation and exploration strategies.

- *defalloc:* The designer can specify the default resource allocation to be either *minimum area* (i.e. complete sharing), *minimum delay* (i.e. no sharing), or *conflict-free* (i.e. the conflict-free allocation of Section 5.2.3). This option is useful if the designer simply wants a logic-level implementation without having to explore the design space.

These settings are used in conjunction with the synthesis commands, described next.

10.2.2 Design subsystems in Hebe

Hebe is organized into a number of *design subsystems.* Each design subsystem performs a particular synthesis task, such as timing constraint specification or design space exploration. We describe now the topmost level of Hebe, called the *system level.*

At the system level, the designer reads in one or more SIF models. For each model that is read in, all models in its calling hierarchy are also recursively read in. Hebe automatically retrieves the design points that were previously synthesized for a given model. The area and delay costs are extracted from these design points and made available in the current synthesis session.

In addition to allowing the SIF models to reside in different directories of a file system, Hebe supports the following two features to assist in the management of design information:

- *Version control:* A design point is tied to the version of the SIF model from which it is generated. To ensure the integrity of the design points if the SIF model is modified by Hercules, the system checks whether the SIF model has been changed after the generation of the design points and notifies the designer accordingly.

- *Incremental synthesis:* Timing and resource constraints, along with the area and delay information for the design points, are physically separated from the SIF and stored as *constraint files.* This means that pertinent information generated during a synthesis session can be check-pointed and retrieved at a later session. This feature supports incremental synthesis where the designer does not have to resynthesize the entire design each time a portion of the design is modified.

A list of the system level commands in Hebe is given in Table 10.3. Details of each design subsystem are presented in the following sections.

System commands	
alias	alias command string to identifier
cd	change current working directory
current	set current model
echo	echo string to terminal
help	print on-line help
history	print history of previous commands
info	print brief tutorial on system
prompt	change system prompt
quit	exit program
set	set default environment variables
Model management commands	
hierarchy	print calling hierarchy for SIF model
printmodel	list SIF models in system
printsif	print SIF model
reset	remove SIF model from system
showsif	graphically display SIF model
summary	print summary of SIF model
Entry to design subsystems	
apply_logic	Logic constraint specification subsystem
apply_resource	Resource constraint specification subsystem
apply_timing	Timing constraint specification subsystem
check_point	Constraint check-point subsystem
explore_space	Design space exploration subsystem
enter_space	Design point synthesis subsystem
str_syn	Perform synthesis with default allocation

Table 10.3: System-level command summary in Hebe Version 3.0.

Constraint specification subsystem

There are three forms of constraints for a particular SIF sequencing model: *logic*, *timing* and *resource* constraints. A design subsystem exists for each type of constraint to allow viewing and editing capabilities of these constraints by the designer. Furthermore, timing and resource constraints can be extracted from a given model and stored as a constraint file, or they can be read from a constraint file and applied to a given model. The *check-point subsystem* is responsible for the interface between these constraint files and the SIF model.

The capabilities of each constraint specification subsystem are described below.

- *Logic constraint subsystem.* A SIF sequencing model consists of a number of *logic expression blocks*, each of which contains combinational logic expressions. To obtain accurate estimates on the area and delay characteristics of these expression blocks, logic synthesis techniques are used. This subsystem allows the logic expressions to be evaluated by writing them as SLIF files and invoking logic synthesis tools. If a library is specified, then the exact area and delay costs can be obtained using technology mapping.

- *Resource constraint subsystem.* The size of the design space depends on the number of calls that have not been bound to resources in a given SIF model. The designer can limit the design space by allocating resources and partially binding operations to specific resource instances. This subsystem provides commands to support these allocation and binding tasks.

- *Timing constraint subsystem.* Timing constraints specify minimum or maximum bounds between the activation of pairs of operations in a given sequencing graph. Delay constraints specify the execution delay of a given operation. Timing constraints can be specified either by the commands of this subsystem, or they can be extracted from the SIF model containing constraints specified in the HardwareC description.

- *Check-point subsystem.* Each SIF sequencing model consists of a set of constraints: *timing*, *resource*, and *dependency*, corresponding to the timing requirements, allocated resources and their bindings, and the sequencing dependencies among the operations. These constraints are incorporated as part of the SIF model. In addition, Hebe maintains a set of *working constraints* for each type of constraint. The working constraints can be extracted from and applied to the SIF model, or read from and written to a given file.

Design space exploration subsystem

The *design space exploration subsystem* allows the designer to generate a set of resource bindings corresponding to a particular resource allocation, and then to evaluate these bindings based on various cost criteria in identifying a set of *candidate bindings.* Each candidate binding is checked to see if the resource conflicts can be resolved under timing constraints. If this is the case, then a *design point* is generated and written to the file system. At this point, the logic-level implementation has not been synthesized. If there are no calls in the model, then its design space trivially reduces to a single design point.

For exact exploration strategy, the complete design space of resource bindings is generated and synthesized, subject to the pruning techniques of Section 5.3.1. For the heuristic exploration strategy, either the complete design space is generated, or if the design space size exceeds the user's limit (e.g. *maxbinding* environment variable) then a portion of the design space is generated. These generated resource bindings are evaluated and ranked according to the following cost criteria: *area*, *interconnect*, and *serialization* costs. The steps in the subsystem are described below.

1. *Set weights and thresholds for the cost criteria.* Weights indicate the relative importance of the cost criteria. Thresholds indicate the limit with respect to a cost criterion. For example, the designer can indicate that only conflict-free bindings are sought by specifying a serialization threshold of 0.

2. *Determine resource allocation.* The designer then specifies a resource allocation for the SIF model. The allocation can be retrieved from a given file or written to a given file for later retrieval. To illustrate what the designer will see, Figure 10.4 shows Hebe's screen display for the *encoder* process example of Section 2.1.5. It indicates that 3 instances of *PARITY(3)* have been allocated out of a maximum of 6, and 1 instance of *PARITY(4)* has been allocated out of a maximum of 3.

3. *Generate resource bindings.* The resource bindings corresponding to the given resource allocation are generated and ranked according to the cost criteria, subject to the appropriate weights and thresholds. The designer can specify that only the top num bindings in the ranking be generated. For example, the top 2 bindings for the *encoder* example corresponding to the allocation of Figure 10.4 are shown in Figure 10.5. For the first binding, `(1 6)(3 4)(2 5)` means that the first and sixth calls to

```
Resource Usage Allocation for model ''encoder'':

 1. PARITY(3)                Number: 6
    Allocation:   [1 2 3]
 2. PARITY(4)                Number: 3
    Allocation:   [1]

Criteria                     Weights            Thresholds
--------                     -------            ----------
Area                         10                 none
Delay                        100                none
Interconnect                 5                  none

Explore: encoder> _
```

Figure 10.4: Screen display in the design space exploration subsystem.

PARITY(3) are implemented by one instance, and the third and fourth calls are implemented by another instance, etc. The area cost (`Area = 18`) means that the three instances of *PARITY(3)* have a combined area cost of 18.

4. *Create design points.* Conflict resolution is performed on the generated bindings. If it is not possible to resolve the resource conflicts subject to the imposed timing constraints, then the binding is considered to be invalid and is discarded. Otherwise, the designer creates one or more design points among the valid bindings.

Each design point represents a particular resource binding configuration for which a valid conflict resolution exists. These points are written to the file system so that at a later session with Hebe, they can be retrieved and presented to the designer for synthesis.

Design point synthesis subsystem

The design space of a SIF model consists of a set of design points. The task of the *design point synthesis subsystem* is to synthesize each design point to generate one or more logic-level implementations. The designer can also simulate the logic-level design and invoke logic synthesis to map the design to different target

```
Design Space of Resource Bindings for ''encoder'':

1. Cost = 675: 10*(Area=27)+100*(Delay=3)+5*(Intercnt=21)
        Resource PARITY(3):   (1 6 )(3 4 )(2 5 )
                              (Area = 18)(D = 3)(I = 9)
        Resource PARITY(4):   (1 2 3)
                              (Area = 9)(D = 0)(I = 12)

2. Cost = 675: 10*(Area=27)+100*(Delay=3)+5*(Intercnt=21)
        Resource PARITY(3):   (2 6 )(3 4 )(1 5 )
                              (Area = 18)(D = 3)(I = 9)
        Resource PARITY(4):   (1 2 3)
                              (Area = 9)(D = 0)(I = 12)

Explore: encoder> _
```

Figure 10.5: Top 2 resource bindings for the allocation of the previous example.

libraries. The exact area and delay characteristics are extracted and stored in a file for later retrieval and update.

10.3 Summary of System Implementation

Hercules and Hebe form a vertically integrated framework for synthesizing synchronous digital designs, with specific attention to the requirements of ASIC designs. The input to the high-level synthesis system is a description of hardware behavior in HardwareC. The resulting implementation is a logic-level specification of logic gates and edge-triggered registers. Hercules and Hebe provide a flexible user interface that supports user customization and version control. The novel algorithms and formulation that are the major contributions of this research, such as relative scheduling, are incorporated as part of the Hebe.

Chapter 11

Experimental Results

This chapter describes the experimental results and design experiences in using the Hercules and Hebe high-level synthesis programs. Several digital ASIC designs were synthesized using this system, including an Ethernet coprocessor [GC91], a Digital Audio input output (DAIO) chip [LBMG89], a bi-dimensional discrete cosine transform (BDCT) chip [RM89], a decoder chip for the space telescope [Kas89], a raster line drawing design, an error-correcting code design, and a greatest common divisor design. Each design was described completely in HardwareC and synthesized to a logic-level implementation. Extensive logic-level simulation demonstrated the correctness of the specification and implementation.

Many of these designs are highly complex. For example, the Ethernet coprocessor is modeled with 13 processes and has over 1200 lines of HardwareC code. Similarly, the DAIO chip contains 2 processes, each of which has over 300 lines of HardwareC code. In addition, the system has been applied to the synthesis of benchmark circuits from the High-level Synthesis Workshop [Wor89]. HardwareC descriptions of these benchmark circuits have been included as part of the official workshop benchmark suite. Although detailed timing constraints and external synchronizations are not part of the specification for many of these benchmark examples, the examples serve to demonstrate the use of this system on general-purpose synchronous digital designs.

This chapter is organized as follows. Section 11.1 presents an overview of the synthesized ASIC chip designs. Section 11.2 describes the synthesis results for several benchmark designs.

11.1 Design Experiences

This section summarizes the design experiences of several ASIC design efforts using Hercules and Hebe at Stanford University. For each design, details of its functionality, system-level requirements, architecture, and implementation statistics are presented. The logic-level implementation consisting of control and data-path is mapped to technology libraries, such as the Actel FPGA library or the LSI Logic gate array library, using the *Ceres* technology mapper [MM90]. For example, the decoder chip for the Multi-Anode Microchannel Array (MAMA) detector that is used in the space telescope [Kas89] was synthesized and fabricated with LSI Logic's 9K-series sea-of-gates technology. All of these designs have been fully simulated to verify their functional correctness.

11.1.1 Ethernet co-processor

The *Ethernet co-processor* [GC91] manages transmitting and receiving data frames over a network under CSMA/CD protocol. CSMA/CD refers to the Carrier Sense Multiple Access with Collision Detection protocol [Tan81], a protocol which is used to facilitate communication among many stations over a shared channel. Any station wishing to transmit listens first to the line and defers its transmission until the channel is clear. Multiple stations are allowed to simultaneously access the channel without use of any central arbitration. The protocol is defined by the IEEE 802.3 standard.

The purpose of the co-processor is to off-load the host processor from managing the communication activities. The co-processor must handle data framing and de-framing, network and link operations, address sensing, error detection, data encoding, and memory access. The co-processor interacts with the host CPU and the main memory through an 8-bit address/data bus. The bus may be controlled by either the co-processor or the host CPU, where control of the bus is arbitrated by means of handshaking signals between the two modules. The co-processor also connects to a local memory where the incoming data frames are stored. A block diagram of the co-processor is shown in Figure 11.1.

The host CPU invokes the Ethernet co-processor by writing to a memory mapped address. The co-processor responds by making a request for control of the address/data bus. When bus control is granted, the co-processor initiates a memory read operation to receive command operations and then releases control of the bus. The co-processor provides a small execution unit with four instructions that let the host processor program the co-processor for specific operations, e.g. transmit some data from memory. The unit also gives the co-processor

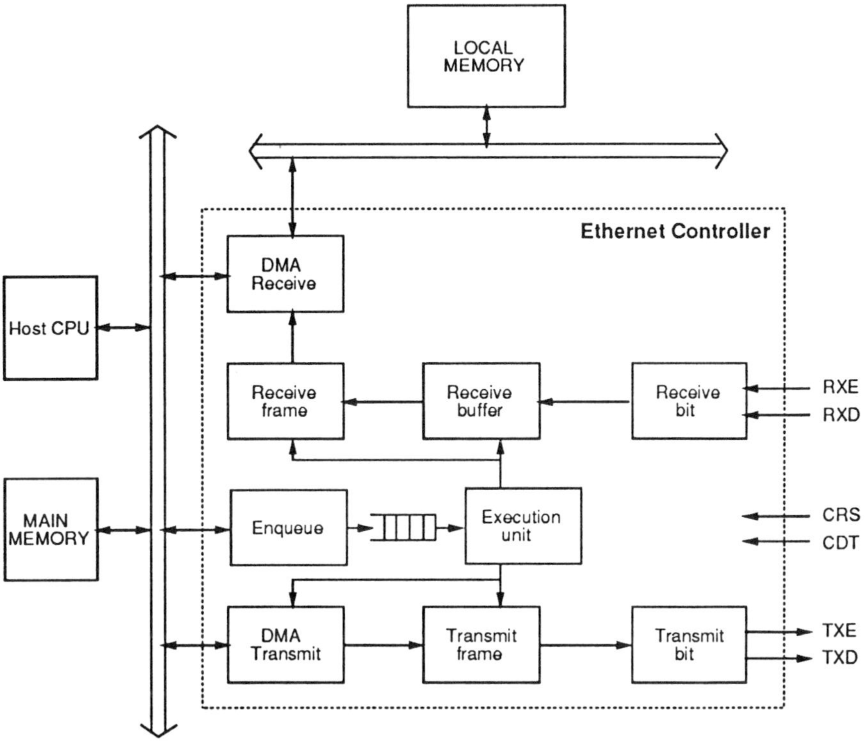

Figure 11.1: Block diagram of the Ethernet co-processor.

some degree of programmability, e.g. configuring the transmission parameters. A commercial co-processor chip with similar functionality is available from Intel [Int87].

The co-processor contains two independent 16-byte wide receive and transmit FIFO queues. Direct memory access (DMA) is supported to off-load the processor during reception and transmission. To receive data frames from the Ethernet line, the host CPU reserves a buffer space in main memory and invokes the co-processor. The asynchronously arriving data frames are received by the co-processor and stored into a local memory. A DMA receiver module then writes the data frames to main memory. To transmit data on the Ethernet line, the host CPU invokes the co-processor with a pointer to the location in main memory containing the data. The co-processor constructs a data frame by reading the data fields from main memory and then transmits it over the Ethernet

Model	*Type*	*Ports*	*Channels*	*lines*
controller	block	39	0	91
exec_unit	process	125	17	140
xmit_bit	process	3	8	46
xmit_frame	process	84	16	217
DMA_xmit	process	55	8	165
DMA_rcvd	process	49	0	80
rcvd_bit	process	12	0	67
rcvd_buffer	process	12	8	20
rcvd_frame	process	54	8	131

Table 11.1: HardwareC description statistics for the Ethernet co-processor.

line. In the event of a collision, the co-processor "jams" for a period of time, performing random wait and retry without CPU intervention.

Hardware description. The entire design is modeled by 13 concurrent processes, described in over 1200 lines of HardwareC code. Table 11.1 summarizes the statistics of the major modules in the co-processor. The model type, number of ports, number of message passing channels, and number of lines of HardwareC code for each module are given. Table 11.2 shows the statistics of the SIF model which is generated for each process. The column named "graphs" represents the number of sequencing graphs in the control-flow hierarchy of the corresponding process. Port read and write commands, as well as message passing send/receive commands, are used extensively in the design description for handshaking purposes.

Block *controller* is the top-level model containing an interconnection of processes. Process *exec_unit* is the execution unit that controls the transmission and reception of data frames. Processes *DMA_rcvd* and *DMA_xmit* are responsible for communicating with the main memory through the shared bus. Processes *recv_frame* and *xmit_frame* are responsible for receiving and sending data frames on the Ethernet line. These two processes are central to the co-processor.

Synthesis results. The Ethernet co-processor has been synthesized using Hercules and Hebe assuming a cycletime of 50 nanoseconds. The logic-level implementation was mapped to the LSI Logic's LCA10K library [LSI87] using *Ceres* [MM90]. Results of synthesis are given in Table 11.3.

Process	*Graphs*	*Nodes*					
		Total	*Cond*	*Loop*	*Send*	*Receive*	*I/O*
exec_unit	13	198	46	13	0	16	26
xmit_bit	2	46	7	2	0	2	19
xmit_frame	17	249	58	10	14	10	33
DMA_xmit	17	215	56	13	7	0	26
DMA_rcvd	4	50	14	1	0	0	8
rcvd_bit	1	42	11	1	0	0	6
rcvd_buffer	4	43	13	4	1	0	3
rcvd_frame	5	111	23	3	0	11	35

Table 11.2: SIF model statistics for the Ethernet co-processor.

The extensive use of external synchronization in the design implies that the length of the schedule cannot be statically defined. We therefore estimate the scheduling cost by two measures: $|A|$ which represents the total number of anchors in the SIF model, and $\sum \sigma_a^{max}$ which represents the sum of the maximum offsets for all anchors in the graphs. The relatively large number of anchors is due to the hierarchical nature of SIF, since the source vertex of every graph in the control-flow hierarchy of a model is also an anchor. If there are no unbounded delay operations and no control-flow hierarchy in the model, then $\sum \sigma_a^{max}$ reduces to the latency of the schedule. Otherwise, it represents the sum of the schedule lengths corresponding to each anchor in the model.

The synthesized design is further characterized in terms of register usage and mapped implementation costs. Register usage statistics consist of the total number of registers and the number of registers due to control and I/O ports. Mapped implementation statistics are based on the LCA10K library parameters. They include the area cost, number of cells, and the critical path delay in nanoseconds. The worst case delay is $45.06ns$, which is within the $50ns$ cycle time bound.

A process marked with an asterisk ($*$) means that technology mapping was performed on the *unoptimized* logic description because the MisII [BRSVW87] logic optimizer ran out of memory. Otherwise, the implementation cost for both the unoptimized and optimized logic description is given. In most cases, combinational logic synthesis significantly reduces the area cost. For example, the area costs for the unoptimized and optimized DMA_rcvd are 1545 and 562, respectively.

The co-processor took approximately one man-month to specify and simu-

process	*Schedule*		*Registers*		*Implementation cost*		
	$\|A\|$	$\sum \sigma_a^{max}$	Total	Ctr/IO	Area	Cells	Delay
exec_unit*	88	42	293	128	5670	2488	23.80 ns
xmit_bit	13	20	32	32	813/352	353/271	14.31 ns
xmit_frame*	107	45	157	142	6310	3183	37.15 ns
DMA_xmit*	93	22	149	109	5902	2560	45.06 ns
DMA_rcvd	19	7	44	23	1545/562	757/400	27.51 ns
rcvd_bit*	12	10	38	16	691	282	12.30 ns
rcvd_buffer	22	3	25	24	366/160	161/127	22.09 ns
rcvd_frame*	37	25	93	56	3198	1571	38.12 ns

Table 11.3: Synthesis results for the Ethernet co-processor.

late. The actual synthesis took less than a few minutes for each process. This is to be compared against an estimated 6 man-months for manual designs of this complexity. C. Coelho and R. Gupta were responsible for the design.

11.1.2 Digital audio I/O chip

The *Digital Audio Input Output* (DAIO) [LBMG89] is a chip which controls the transfer of data between a microprocessor and a compact disc player or a digital audio tape player. It converts a serial line interface protocol defined by the Audio Engineering Society (AES) to a standard 16/32-bit microprocessor bus interface. The AES protocol specifies a format for the bit-serial synchronous transmission of two channels of audio signals. Data is transmitted in blocks, each containing 192 frames. Each frame consists of two 32-bit subframes, one for each audio sample. A block diagram of the DAIO and its interface are shown in Figure 11.2.

The DAIO is a full duplex, fully symmetrical interface chip for the reception and transmission of digital audio signals. Each direction, receive and transmit, has its own set of dedicated 32-bit registers. The register bank holds a total of 4 frames for each direction and is double buffered. The host processor controls the DAIO's operation by writing to a set of *mode* and *status* registers.

Synthesis results. Two versions of the DAIO chip were designed at Stanford University. The first version, designed by M. Ligthart, consisted of only the receiver and phase decoding features. The implementation was synthesized and

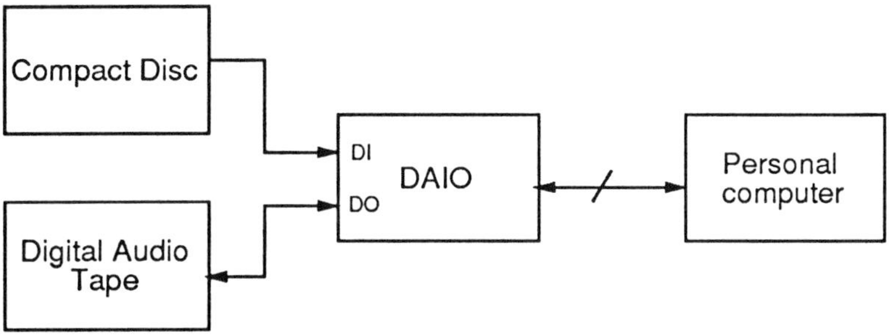

Figure 11.2: Block diagram of a Digital audio I/O (DAIO) chip.

process	*HardwareC*	*SIF model*		*Implementation cost*			
		Nodes	Graphs	Regs	Area	Cells	Delay
daio_receiver2	308	64	16	52	697	498	38.80 ns
phase_decoder2	323	58	9	27	566	404	20.35 ns

Table 11.4: Synthesis results for the second version of the DAIO chip.

mapped onto LSI Logic's macrocell library for the LCA10K Compacted Array series [LSI87]. We refer the interested readers to the publication authored by Ligthart [LBMG89] for details of the synthesis results. The second version re-implemented DAIO's receiver and phase decoder section using a slightly different design approach to achieve better performance. The two processes, *phase_decoder2* and *daio_receiver2*, were again mapped to an implementation in LSI Logic's LCA10K library. R. Gupta and T. Truong were responsible for this revision.

Table 11.4 summarizes the statistics of both the SIF model and the mapped implementation. For each process, the total number of nodes and graphs in the SIF model and the number of lines of HardwareC code are given. Characterization of the implementation is in terms of the number of registers, the area cost, the number of library cells used, and the critical path delay in nanoseconds. The synthesized design can execute with a clock rate of 25 MHz.

Figure 11.3 shows a simulation trace of the DAIO phase decoder using the *Mercury* logic-level simulator [MKMT90]. The phase decoder monitors the biphased coded data stream `data` from the audio device and extracts the following synchronization signals: `preamble_1`, `preamble_2`, and `preamble_3`.

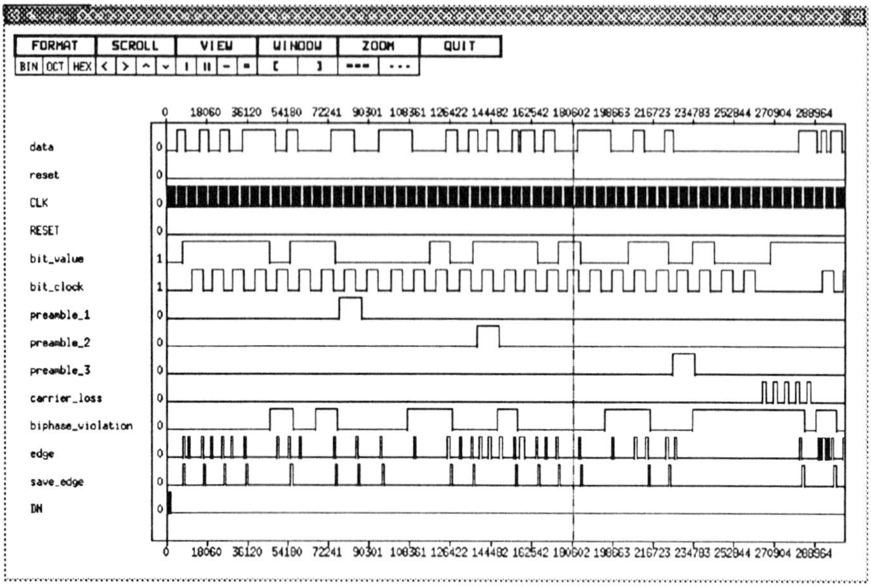

Figure 11.3: Simulation trace for the DAIO phase decoder.

When it detects the occurrence of biphase violations (transitions between the bits), the signal `biphase_violation` is asserted. The detected edge of the data stream is indicated by `save_edge`. The phase decoder also derives the clock and bit value in `bit_clock` and `bit_value`, respectively.

11.1.3 Bidimensional discrete cosine transform

The *Bi-dimensional Discrete Cosine Transform* (BDCT) [RM89] is a 8×8 video rate coding chip that is used to remove redundant video information in low bit-rate transmission channels and to perform video compression for image storage and retrieval. The architecture is motivated by the observation that the BDCT equations can be solved row-by-row and column-by-column by a simpler Mono-dimensional DCT (MDCT). The design is therefore partitioned into three stages: the first and last stage implement MDCTs while the second stage is a shared memory array. The intermediate stage is a storage array for a matrix, implemented as a double threaded shift register to allow transposition of the matrix. The control unit coordinates the activities of these three stages. A block diagram of the BDCT is shown in Figure 11.4.

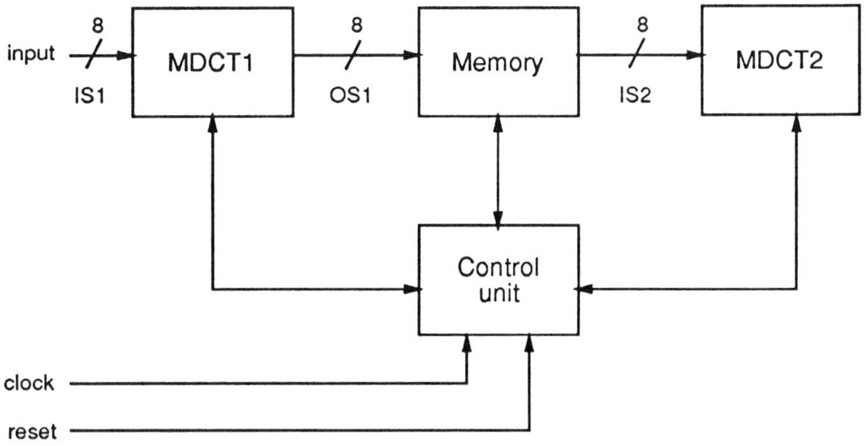

Figure 11.4: Block diagram for the Bi-dimensional DCT chip.

Hardware description. The BDCT chip was described in HardwareC and automatically synthesized by Hercules[1]. The design consists of a set of concurrent processes that communicate through a shared medium (Memory module in Figure 11.4). The shared memory is not described in HardwareC, but instead is mapped by a module generator directly to layout. Each MDCT module is described as two concurrent processes (*phase-A* and *phase-B*). Process *phase-A* computes the additions (15) and subtractions (16) in parallel. The output of process *phase-A* is fed to process *phase-B* which performs the serial multiplications in parallel. The description of the two MDCT modules is identical except for the computation parameters. A full description of the MDCT in HardwareC is reported in [RM88]. The circuit was simulated at the functional and circuit level using Lsim [Sys88].

Synthesis results. The physical layout of the BDCT chip is obtained by interconnecting macro-cells, where each macro-cell is automatically generated by the *Castor* and *Pollux* programs [MM88]. *Castor* takes as input the SLIF description and produces a symbolic layout. This symbolic layout is then transformed by *Pollux* into a geometric layout based on the GDT design rules [Sys88]. The chip was designed using a 2.0μ technology. Table 11.5 summarizes the final implementation. The core size is 6.9 by 7.6 mm and the die size 9.5 by 9.2 mm.

[1]BDCT was designed using an earlier version of Hercules and HardwareC.

Stage	Process	Size	Rows	Transistor	Cells
$MDCT_1$	*phase-A*	$900 \times 1250\mu^2$	12	832	352
	phase-B	$2900 \times 3700\mu^2$	24	4634	2105
$MDCT_2$	*phase-A*	$900 \times 1250\mu^2$	12	832	352
	phase-B	$3150 \times 3850\mu^2$	24	6050	2213
CONTROL	*control*	$1100 \times 1050\mu^2$	12	954	392

Table 11.5: Synthesis results for the Bi-dimensional DCT chip.

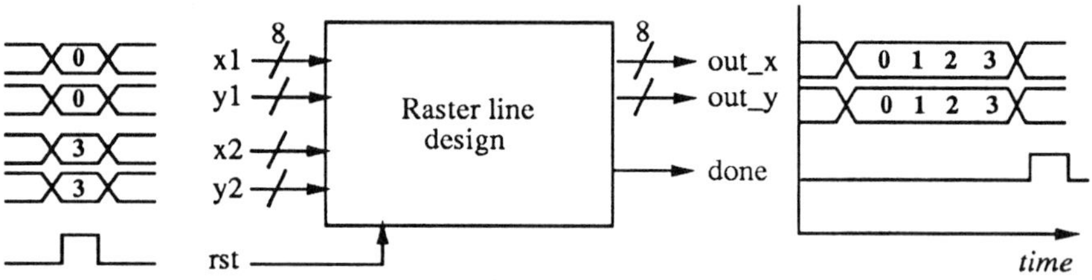

Figure 11.5: Block diagram for the raster line design.

There are 20 macrocells, totaling to 8904 gates excluding the I/O pads. There are 12 input, 8 output and 14 tristate bidirectional pads, with 8 power/ground pads. The chip has been designed to operate at a frequency $F_p = 13.5$ MHz to be compatible with standard data communication rates although higher rates can be sustained. V. Rampa was responsible for the design of the BDCT chip.

11.1.4 Raster line design

The *Raster line* design implements an efficient line drawing algorithm due to Bresenham [HB86] to determine the pixel positions in a graphic display system. This algorithm finds the closest integer coordinates to the actual line path using only integer arithmetic. A block diagram of the raster line design is shown in Figure 11.5.

Hardware description. The raster line design is described in a HardwareC process called *raster*. Taking as input the x and y coordinates of the two endpoints, the design produces a sequence of pixel positions corresponding to the line connecting the two endpoints. Coordinates are 8-bit quantities. Inputs are sampled on the rising edge of the `rst` signal; the completion of the output

Process	*HardwareC*	*SIF model*					
		Graphs	*Nodes*	*Cond*	*Loop*	*Calls*	*I/O*
Raster	100	32	139	14	3	24	6

Table 11.6: SIF model statistics for the raster line design.

Design	*Implementation cost*			
	Schedule	*Register*	*Area*	*Delay*
rip_sub_8	0	0	112	15.70 ns
rip_add_8	0	0	80	14.90 ns
raster				
(1+1-)	31	224	5437	49.27 ns
(1+8-)	23	192	5110	49.27 ns
(3+2-)	26	216	5536	66.22 ns
(16+1-)	17	128	4708	59.51 ns
(16+8-)	7	96	4173	60.01 ns

Table 11.7: Synthesis results for the raster line design.

sequence is indicated by the assertion of the done signal. Process *raster* calls an 8-bit addition *rip_add_8* 16 times and an 8-bit subtraction *rip_sub_8* 8 times. Table 11.6 summarizes the SIF model statistics for the design.

Synthesis results. The final design was mapped to an implementation in LSI Logic's LCA10K Compacted Array library with a cycle time of 67 nanoseconds to sustain a 15 MHz clock frequency. Implementation statistics include the number of registers, the area cost using library parameters, and the critical path delay in nanoseconds. As in the Ethernet co-processor design, we estimate the scheduling cost as the sum of the maximum offsets ($\sum \sigma_a^{max}$) in the relative schedule for all graphs in the SIF model, since the schedule latency cannot be statically derived due to the presence of data-dependent loops in the description.

We explored several points in the design space and their synthesis results are summarized in Table 11.7. The first design point, $(1+1-)$, allocates the minimum amount of hardware. The corresponding schedule cost of 31 is the maximum among all design points. The second design point, $(1+8-)$, allocates one adder and dedicates hardware for each of the 8 subtractions. In this case,

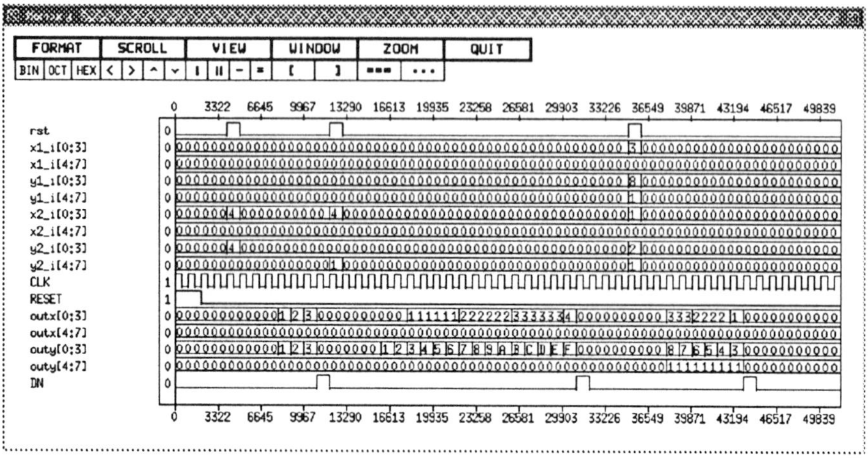

Figure 11.6: Simulation trace for the raster line design.

the schedule cost is decreased to 23 because the extra cycle required to latch a shared subtractor is now eliminated. The third design point, $(3+2-)$, reduces the allocation to 3 adders and 2 subtractors. However, the resulting area and schedule costs are increased due to the increase in the number of multiplexers and latching registers. The fourth design point, $(16+1-)$, dedicates hardware for each of the 16 additions and allocates a single subtractor. The area and schedule costs decrease, because multiplexers and latches are not needed to implement the additions. The final design point, $(16+8-)$, completely dedicates hardware for each addition and subtraction. In this case, the reduction in multiplexing and latching costs is much more significant.

Since *rip_add_8* and *rip_sub_8* are fairly small in area, the benefits of sharing these modules in the raster design are balanced by the corresponding increase in multiplexing and latching costs. Therefore, dedicating hardware resources for each operation produces the minimum area and schedule cost implementation for this design. A simulation trace for the implementation corresponding to the final design point is shown in Figure 11.6.

11.1.5 Error correcting code

The *Error correcting code* (*ECC*) design simulates the serial transmission of data on a noisy line with single-bit error correction. The design consists of three processes, *encoder*, *decoder*, and *noise*. On the rising edge of `new_data`,

Model	*Graphs*	*Nodes*					*lines of SIF*
		Total	*Cond*	*Loop*	*Call*	*I/O*	
PARITY_3	1	3	0	0	0	0	31
PARITY_4	1	3	0	0	0	0	32
encoder	4	47	1	1	9	19	482
decoder	4	53	1	1	9	20	706

Table 11.8: SIF model statistics for the Error correcting code design.

an 8-bit bus `data_in` is sampled by the encoder. The encoder computes the parity bits and serially transmits the 16-bit data frame (8-bit data + 8 parity bits) on the serial line `encoder_out`. Transmission of the data frame is prefixed by the rising edge of the `data_ready` signal. The noisy line is simulated by inverting the value on the serial line whenever the `error` signal is asserted. The decoder samples the data on this noisy line and performs single-bit error correction to recover the original parallel data. The extracted data and the status of transmission (i.e. whether error has occurred) are written out on the `data_out` bus and the `detected_error` signal, respectively. Validity of the decoder output is indicated by the assertion of `out_ready` signal.

To compute the parity bits, the encoder calls a 3-bit *PARITY* function (PARITY_3) 6 times and a 4-bit *PARITY* function (PARITY_4) 3 times. Likewise, the decoder computes the parity bits by calling the PARITY_4 function 9 times. Section 2.1.5 described in detail the HardwareC specification for *ECC*. A block diagram of the design was shown in Figure 2.8. Table 11.8 summarizes the statistics on the SIF model of each process.

Synthesis results. The *ECC* was synthesized and mapped to cells in LSI Logic's LCA10K library. Two experiments were performed to illustrate the effect of resource sharing. In the first experiment, the hardware resources implementing PARITY_3 and PARITY_4 were assumed to have small area cost. This assumption corresponds to the actual implementation of these resources based on their HardwareC description, e.g. PARITY_3 is a three bit XOR and PARITY_4 is a four bit XOR. In the second experiment, the area cost for these resources was increased ten-fold to demonstrate the case where the area reduction due to sharing resources outweighs the area increase due to multiplexers and latching registers. Synthesis results of these experiments are summarized in Table 11.9.

A number of design points were explored and synthesized. For the encoder,

Model	*Resources*		*Implementation cost*			
	PARITY_3	PARITY_4	*Schedule*	*Register*	*Area*	*Delay*
PARITY_3			0	0	6	2.30 ns
PARITY_4			0	0	9	3.50 ns
encoder	1	1	23	45	603	10.40 ns
	2	1	20	42	582	10.40 ns
	6	1	20	36	504	10.40 ns
	6	3	17	323	452	14.30 ns
decoder	-	1	27	59	886	12.90 ns
	-	2	23	55	862	13.10 ns
	-	7	21	53	810	14.84 ns
	-	9	18	50	770	17.43 ns

(a) Implementation assuming cheap resources (actual).

Model	*Resources*		*Implementation cost*		
	PARITY_3	PARITY_4	*Schedule*	*Register*	*Area*
PARITY_3			0	0	60
PARITY_4			0	0	90
encoder	1	1	23	45	738
	2	1	20	42	794
	6	1	20	36	909
	6	3	17	323	1019
decoder	-	1	27	59	967
	-	2	23	55	1024
	-	7	21	53	1377
	-	9	18	50	1499

(b) Implementation assuming expensive resources (hypothetical).

Table 11.9: Synthesis results for the Error correcting code design.

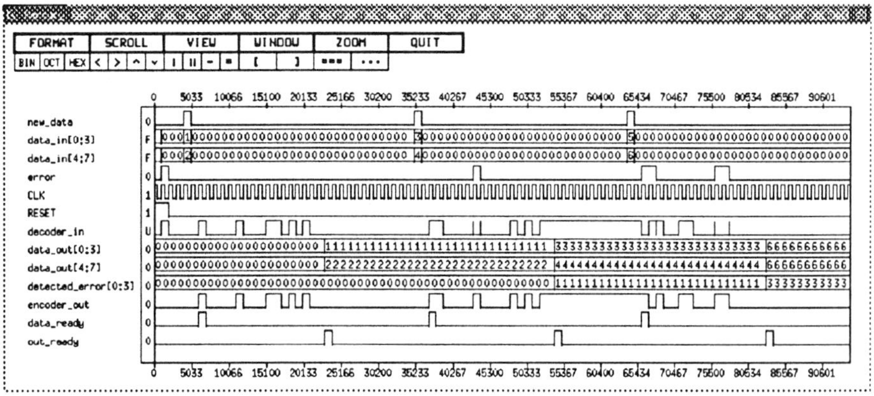

Figure 11.7: Simulation trace for the error correcting code design.

the following allocations were explored: (1,1), (2,1), (6,1), and (6,3), where the first number in the parenthesis represents the allocation for PARITY_3 and the second number represents the allocation for PARITY_4. For the decoder, the explored allocations for PARITY_4 include 1, 2, 7, and 9. The conflict-free allocation for the encoder corresponds to 6 instances of PARITY_3 and one instance of PARITY_4. For the decoder it corresponds to 7 instances of PARITY_4.

The implementation for each design point is characterized in terms of the schedule cost, number of registers, area cost, and critical path delay. The number of registers decreases as more resources are dedicated because it is not necessary to latch the output of shared resources. The critical path delay increases as more resources are dedicated due to the effect of cascading logic.

In the first experiment, the area reduction achieved by resource sharing is greatly exceeded by the increase in the number of multiplexers and latching registers. For example, the area of the minimum allocation encoder implementation is 25% greater than that of its dedicated hardware implementation. For the second experiment, the schedule cost and the number of registers and multiplexers remain the same as before. However, since the area cost of a resource is increased ten-fold, the benefits of resource sharing is more significant. For example, the area of the minimum allocation encoder implementation is 25% less than the dedicated hardware implementation.

Figure 11.7 shows a simulation trace of the error correcting design assuming a dedicated hardware implementation. The first sample of `data_in` is `0x21`. The decoder correctly extracts the original data because no error occurred during

Expr	*Unmapped logic*		*Actel costs*		*LSI costs*	
	Literals	Levels	Area	Delay	Area	Delay
1	1	0	1	9.0 ns	1	1.0 ns
2	2	0	0	0.0 ns	0	0.0 ns
3	37	9	5	19.0 ns	10	3.0 ns
4	179	17	39	134.0 ns	69	11.0 ns
5	438	22	91	197.0 ns	188	22.0 ns
6	16	7	3	19.0 ns	6	3.0 ns
7	138	2	32	19.0 ns	64	1.0 ns

Table 11.10: Summary of expression blocks in the Greatest common divisor design.

transmission. The second sample of `data_in` is `0x43`. Although an error occurred during transmission, the decoder is able to perform error correction to extract the correct data. The status `detected_error` is set to 1 to indicate that a single-bit transmission error was detected and corrected. On the final sample, multiple errors occurred during transmission; the decoder is not able to retrieve the original data and sets the status signal accordingly.

11.1.6 Greatest common divisor

The *greatest common divisor* (*gcd*) design computes the greatest common divisor of two 8-bit numbers using Euclid's algorithm. The HardwareC description for *gcd* was described in Figure 2.2 of Chapter 2. We briefly review its functionality. On the rising edge of the `rst` signal, *Gcd* samples two 8-bit input busses `xi` and `yi`. It then performs Euclid's iterative algorithm to compute the greatest common divisor. The result is written out to the `result` bus along with a pulse in the `DN` signal.

Synthesis results. We use the *gcd* design to illustrate the extraction of area and delay information from logic synthesis. The area and delay estimates are based on both the Actel library and LSI Logic's LCA10K library. The extracted information for each of the seven expression blocks in *GCD* is summarized in Table 11.10. The column named "unmapped logic" gives the number of logic literals and logic levels in the unmapped logic expressions. The Actel and LSI costs describe the area and delay characteristics of an expression block when

Library	*Schedule*	*Register*	*Eq. Gates*	*Area*
Actel	9	35	855	430
LSI LCA10K	6	32	843	1111

Table 11.11: Synthesis results for the Greatest common divisor design, assuming 100ns cycle time.

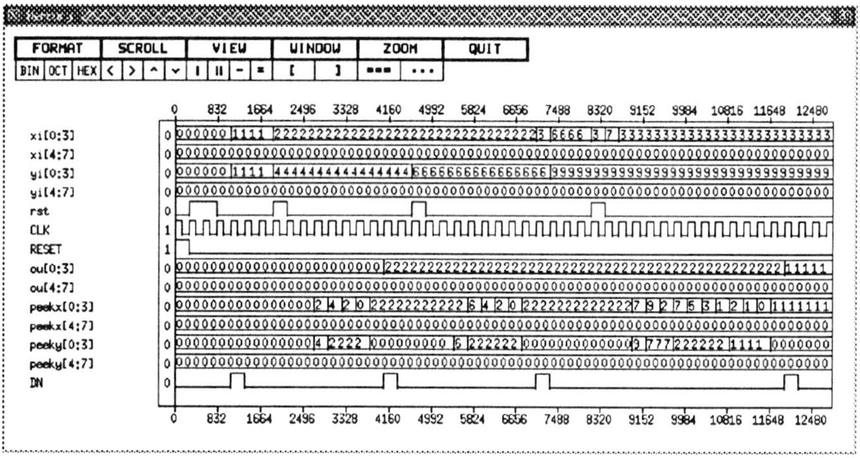

Figure 11.8: Simulation trace for the greatest common divisor design.

mapped to the given library. For example, expression block #5 has an area cost of 188 and a critical path delay of 22.0 nanoseconds when computed using the LSI library. Expression block #2 has zero delay and cost because its logic is reduced to direct wiring.

The synthesis results for *gcd* are summarized in Table 11.11. The design has been mapped to both Actel and LSI Logic's LCA10K libraries with a cycle time of 100ns. The schedule cost is the sum of the maximum offsets of the relative schedule. The number of equivalent gates corresponds to the number of 2-input NANDs/NORs required to implement the circuit. Note that because of expression blocks #4 and #5, the schedule cost for the Actel implementation is greater than the LSI implementation.

Figure 11.8 shows a logic-level simulation trace for *gcd*. The progress of Euclid's iterative algorithm is illustrated by the `peekx` and `peeky` signals. Detailed timing constraints are imposed on the sampling of the inputs as follows:

```
<
    constraint mintime from tagA to tagB = 1 cycles;
    constraint maxtime from tagA to tagB = 1 cycles;

    tagA:  y = read(yi);
    tagB:  x = read(xi);
>
```

Input `xi` is to be sampled *exactly* one cycle after the sampling of `yi`. The effect of timing constraints is illustrated in the third sample of the simulation trace. With the exception of the value 7 appearing exactly one cycle after the rising edge of `rst`, the input `xi` is maintained at the value 3. The design correctly samples 7 and 9 from the inputs and produces the result gcd(7,9) = 1. If timing constraints were not enforced in the synthesized design, then *gcd* would have computed instead the greatest common divisor of 3 and 9, the result of which is 3.

11.2 Synthesis of Benchmark Examples

Hercules and Hebe have been applied to the design examples in the benchmark suite proposed by the High-level Synthesis Workshop [Wor89]. Most of these benchmark examples have been rewritten in HardwareC and synthesized to logic-level implementations.

Three widely used and compared examples are the example used in Facet (Tseng) [TS86], the differential equation solver (Diffeq) [PKG86], and the 5th order elliptic waveform filter (Elliptic) [DDN85]. Although these examples do not contain detailed synchronization and timing constraints, they serve to demonstrate the use of our approach on general synchronous designs. The statistics on the SIF models of these three benchmark designs are given in Table 11.12. The size of the multiplication is double the size of the addition in these examples. For example, Elliptic requires 32-bit multiplications and 16-bit additions.

To evaluate and compare the results of resource binding and scheduling in Hebe with existing systems, we make the following assumptions:

- Additions (both 8-bit and 16-bit) and subtractions take 1 cycle to execute.
- Multiplications (both 8-bit and 16-bit) and divisions take 2 cycles to execute. Multiplication is non-pipelined.
- No detailed timing constraints are present.

Design	Lines		*Nodes*	*Graphs*	*Calls*				*Size*
	HC	SIF			add	sub	mul	div	
Tseng	79	256	11	1	3	1	1	1	8/16-bit
Diffeq	81	196	17	2	2	2	6	none	8/16-bit
Elliptic	146	390	37	1	26	none	8	none	16/32-bit

Table 11.12: SIF model statistics for the benchmark examples.

Resources		*Latency (cycles)*					
add_{16}	mul_{32}	FDS	FDLS	CSTEP	SLICER	PBS	*Hebe*
1	1	–	–	–	–	–	28
2	2	19	18	19	19	18	19
3	1	–	–	–	–	–	23
3	2	18	–	–	18	18	18
3	3	17	17	17	–	17	17

Table 11.13: Comparison of schedule latency for the Elliptic filter example.

For the elliptic filter example, the filter coefficients are arbitrary 16-bit values and not necessarily powers of two. Therefore, multiplication with these coefficients are implemented as a full multiplication with a set of 16-bit wide coefficient registers instead of shift registers. Our results are compared with the force-directed scheduling (FDS) and force-directed list scheduling (FDLS) in HAL [PK89a], the CSTEP scheduler (CSTEP) [TDW+90], SLICER in Chippe (SLICER) [PG87], percolation based scheduling (PBS) [PLNG90]. These results are based on the original Elliptic specification that is distributed in the benchmark suite. Tables 11.13 summarize the comparison of the schedule latency for different resource allocations in the Elliptic example. The schedule latency is known because there are no unbounded delay operations in the design. A dash (–) in the table means that synthesis result of the corresponding synthesis system is not available in the literature.

Tables 11.14 summarizes the synthesis results for Tseng's example and the Diffeq example. Comparison with other systems is not available because the published results are in terms of operator cell area. For example, HAL uses relative operator cell area to describe the area cost [PKG86].

The heuristic design space exploration and conflict resolution strategy was

Design	*Resources*	*Latency*
Tseng	$1+, 1-, 1*, 1/$	4 cycles
	$2+, 1-, 1*, 1/$	3 cycles
	$3+, 1-, 1*, 1/$	3 cycles
Diffeq	$1*, 1+, 1-$	13 cycles
	$1*, 2+, 1-$	13 cycles
	$2*, 1+, 1-$	7 cycles
	$2*, 2+, 1-$	7 cycles
	$3*, 2+, 1-$	6 cycles
	$4*, 2+, 1-$	6 cycles

Table 11.14: Summary of schedule results for Tseng and Diffeq examples.

Module	*Size*	*Latency*	*Register*	*Area*	*Delay*
Add	8 bits	0	0	87	10.95 ns
	16 bits	0	0	188	11.75 ns
Subtract	8 bits	0	0	120	18.50 ns
Multiply	16 bits	0 cycle	0	3382	58.20 ns
		2 cycles	65	2424	43.00 ns
		4 cycles	67	1417	40.20 ns
	32 bits	4 cycles	132	5945	74.70 ns
		20 cycles	215	4698	49.10 ns

Table 11.15: Synthesis results for the arithmetic library modules.

used to obtain these results. Note that the latency can be improved if expression tree height reduction is performed during behavioral synthesis. However, this optimization is not implemented in the current version of Hercules.

Synthesis results. One of the advantages of synthesis approach presented in this book is that the resulting hardware is described as a synchronous logic-level implementation. This logic-level implementation can be operated on by logic synthesis tools and mapped to cells of a given library. We present synthesis results of the Tseng, Diffeq, and Elliptic examples for different allocations of resources. Each implementation is mapped to LSI Logic's LCA10K library using the technology mapper *Ceres*.

Benchmark	*LSI Implementation*			
	Schedule	*Reg*	*Area*	*Delay*
Elliptic $(1*, 1+)$	47	456	9508	69.60 ns
Elliptic $(2*, 1+)$	45	470	9830	69.60 ns
Elliptic $(2*, 2+)$	31	464	11991	69.65 ns
Tseng $(1+, 1-, 1*, 1/)$	6	192	3101	72.40 ns
Tseng $(2+, 1-, 1*, 1/)$	5	183	3058	72.40 ns
Tseng $(3+, 1-, 1*, 1/)$	5	167	2904	77.40 ns
Diffeq $(1+, 1-, 1*)$	31	129	2320	36.10 ns
Diffeq $(1+, 1-, 2*)$	16	129	2626	37.17 ns

Table 11.16: Synthesis results for the benchmark examples, assuming 5 cycle 16-bit and 32-bit multiply (4 cycle multiply + 1 cycle latch).

Instead of assuming the availability of a 2 cycle multiplier from a given micro-architectural library, we have designed and implemented an efficient multiplier unit starting from a HardwareC description. Tradeoffs can be performed on this multiplier just as with other designs. Table 11.15 gives the statistics on the area and delay costs for arithmetic library modules. Note that a 32-bit multiplier taking two 16-bit operands is significantly larger in area than a 16-bit adder. This is contrary to the assumption made by most synthesis systems that a multiplier is 4 times as large as an adder.

Based on these library resources, the implementation results for different allocations of Tseng, Diffeq, and Elliptic are presented in Table 11.16. For Elliptic, the 4 cycle 32-bit multiplier was used to implement its multiplication. For Tseng and Diffeq, the 4-cycle 16-bit multiply was used. Combinational logic optimization was not performed on the logic-level implementation prior to technology mapping.

11.3 Summary of Experimental Results

Hercules and Hebe have been applied to the synthesis of synchronous ASIC designs. These designs are described in HardwareC at the behavioral level and synthesized to logic-level implementations. Logic synthesis is performed to map the logic-level implementation to the cells of a given technology library. Hercules and Hebe have also been applied to the benchmark circuits of the High-level Synthesis Workshop. The synthesis results are comparable with existing approaches.

Most synthesized designs require complex coordination with their environment to meet handshaking and timing protocols. For example, the Ethernet co-processor design demonstrates the extensive use of external synchronization in both specification and synthesis. The underlying hardware model for both HardwareC and Hebe allows these communication requirements to be clearly specified and enforced in the synthesized hardware.

Chapter 12

Conclusions and Future Work

High-level synthesis of VLSI circuit designs offers an effective means to deal with the increasing complexity of digital hardware design. Previous synthesis approaches have primarily been targeting general-purpose processor and signal processing applications. This research focuses instead on the synthesis of ASIC designs. ASIC designs are typified by network controllers and communication modules for which system integration plays a important role in the design process. These designs often have complicated handshaking and timing requirements on their interface with other concurrently executing modules in the system. Because of the complexity of ASIC designs, reducing the design time is as important as minimizing area or improving performance.

We have identified three requirements in the modeling of ASIC designs; they are the support for *concurrency*, *external synchronizations*, and *detailed timing constraints*. The overall goal of this research was to incorporate these requirements into the hardware model and to develop synthesis algorithms that operate on this hardware model. This book has described novel synthesis algorithms and their implementation in a high-level synthesis system. The major highlights are summarized in the following sections.

12.0.1 Summary of Major Results

Synthesis Algorithms. We have developed a *sequencing graph model* to capture the three requirements of ASIC designs. The model supports: concurrency by allowing for multiple threads of concurrent execution flow, external synchronization by modeling both operations with fixed and unbounded execution delays, and detailed timing constraints by specifying minimum and maximum bounds on the activation of operations. The sequencing graph model is the underlying representation for the synthesis formulation and algorithms. Because of its generality, however, existing synthesis approaches cannot be directly applied to this hardware model. For example, the support for unbounded delay operations in the hardware model invalidates traditional scheduling formulations because operations cannot, in general, be statically assigned to control steps.

To incorporate external synchronization and detailed timing constraints, we have developed a novel formulation called *relative scheduling*. Relative scheduling supports the scheduling of both fixed and unbounded delay operations under detailed timing constraints. It provides a theoretical framework for analyzing consistency of timing constraints in the presence of unbounded delays. Polynomial-time algorithms have been developed to analyze consistency of constraints as well as to find a minimum relative schedule. Furthermore, we have formulated the problem of resolving resource conflicts in the graph movel in terms of serializing operations that are bound to the same resource. A technique called *constrained conflict resolution* was developed to resolve conflicts so that the resulting design still satisfies its handshaking and timing requirements. This technique is based on the relative scheduling formulation and exploits the topology of timing constraints to reduce its computation time.

To support the relative scheduling framework, we have developed an approach called *relative control synthesis* to generate the control logic to activate operations according to a given relative schedule. The control logic is synthesized as an interconnection of interacting finite-state machines that coordinate with one another via handshaking signals. In addition to the support for concurrency and external synchronization, the resulting control is guaranteed to execute in the *minimum* number of cycles for all input data sequences. This feature is important to ensure that the control hardware incurs no performance penalties for the arbitrary nesting of model calls, conditionals, and loops in the behavioral description. Relative scheduling and relative control provide a uniform framework for the analysis and synthesis of the sequencing graph model.

Finally, we have developed a novel optimization strategy for control synthesis that operates on the sequencing graph abstraction. This approach, called *relative control optimizations*, departs from more conventional control optimizations that are based on the FSM model in that it takes advantage of the degrees of freedom present at the behavior level. In particular, it minimizes the size of the control implementation subject to the required timing and synchronization constraints.

System Implementation. An aim of this research project was to complement the theoretical development by constructing a practical synthesis tool which embodies the theory and that can be applied to complex ASIC designs. Towards this objective, we have developed a hardware description language called *HardwareC* and a high-level synthesis system called *Hercules* and *Hebe* that are targeted towards synchronous ASIC designs.

HardwareC supports many features that are useful in specifying ASIC designs, such as inter-process synchronization via send/receive message passing, varying degree of parallelism, explicit instantiation of models, and detailed timing constraints. This language serves as the input to Hercules and Hebe, which are the implementations of the algorithms developed in this research. Hercules and Hebe form a vertically integrated synthesis system that incorporates logic synthesis as the basis for logic-level optimization and evaluation. Different design alternatives can be explored in this framework, where the designer is guided by both exact and heuristic search strategies to find an acceptable implementation satisfying the required timing and resource constraints. The system also supports user customization of the interface and version control of the design data.

This system has been applied to the synthesis of several complex ASIC designs, including an Ethernet controller, a digital audio I/O chip, and a bi-dimensional DCT chip. These designs have been described in HardwareC and synthesized to a gate-level netlist. Extensive simulation has been performed to verify the correctness of the specification and the resulting implementation. In addition, the system has been applied to the synthesis of benchmark circuits from the High-level Synthesis Workshop. Based on these design experiences, we have demonstrated that the synthesis approach described in this book is effective in synthesizing complicated ASIC designs.

12.1 Future Work

Computer-aided synthesis of digital circuits from an algorithmic description is a difficult problem. We have proposed in this research an approach to ASIC synthesis that addresses some of the difficulties in effectively integrating a synthesized design with other components in the system. However, many difficult problems have not yet been solved. We summarize in Section 12.1.1 the current limitations of this research and propose avenues of further research in Section 12.1.2.

12.1.1 Current limitations

This research focused only on synchronous and non-pipelined hardware designs. For certain signal processing applications, pipelining can significantly increase the throughput of the design [HJL+89]. It is therefore desirable to extend the relative scheduling and relative control formulations to support pipelining. For designs without external synchronization, such extension is straightforward. For designs with complicated handshaking and timing requirements, however, pipeline flushing and branching are nontrivial issues that need to be addressed.

Asynchronous circuit synthesis, as well as designs with poly-phase clocking strategies, are not considered in this book. Automated synthesis of asynchronous circuits is an active area of research [MMB89, Chu87] and multiphase circuits can be useful for high performance circuits [BBR91]. The integration of these different circuit styles is an important prerequisite in synthesizing complex, multiple chip circuits. Although these circuit styles are not considered in this research, the techniques described in this book can serve as a starting point for such efforts. For example, for multi-phase synthesis, retiming [LRS83, BBR91] can be performed on the resulting logic-level implementation to balance the logic across multiple clock phases.

A limitation of our approach is that timing constraints are applied only on operations within the same sequencing graph. Although it is possible to distribute the constraints across the hierarchy using simple heuristics [Nes87] (i.e. depth-first traversal), thorough analysis of the different tradeoffs in constraint distribution in the presence of resource constraints is still lacking.

Finally, it is sometimes the case that although the execution delay of an operation is not fixed, it is bounded within a certain range of values. For example, an operation may require between 2 and 5 cycles to execute. Currently, this operation is modeled as an unbounded delay operation in relative scheduling. However, consistency analysis assuming this range of execution delays should still be possible. We are currently working on extending relative scheduling to address this issue.

12.1.2 Future directions

Spatial and temporal timing constraint tradeoffs. Timing constraints can be categorized into two types, *spatial* and *temporal*. Spatial timing constraints correspond to minimum and maximum bounds on the activation of operations. These constraints can be used during scheduling to ensure that the resulting control satisfies these requirements. Temporal timing constraints correspond to timing requirements on a sequence of events that passes through a given operation. These constraints may describe throughput requirements, i.e. two consecutive data frames are spaced two cycles apart. Unlike its spatial counterpart, temporal timing constraints cannot in general be directly incorporated into the synthesis algorithms. They are usually manually converted into a set of spatial timing constraints. However, this is an error prone and difficult task because of the intricate interaction that exists among the operations in a given behavior. An analysis of this conversion process and the tradeoffs that are involved is a critical link to effectively capturing constraints from the designer.

System-level synthesis. One of the most challenging and difficult tasks in hardware design is to transfer an abstract description in terms of waveforms, textual descriptions, and complex communication protocols into an algorithmic description suitable for high-level or automated synthesis. System-level partitioning plays an important role in defining the interface and functionality of components in the system. However, the highly inter-dependent nature of these components means that specifying one component in detail tends to impose restrictions on the interface and functionality of the other components in the system. The support for system-level partitioning and architectural synthesis increases in importance as designs become more complex.

Graphical visualization of design process. Graphics play an important role in the human design process. A designer usually starts a design by drawing a block diagram to visualize the system partitioning, then successively refines the block diagram to fill in the details as the design progresses. Instead of using a textual hardware description language such as VHDL or HardwareC, it is perhaps more effective to leverage the graphical visualization of the design process. Graphical visualization allows more effective debugging and evaluation of the design.

System-level modeling for synthesis. With few exceptions, existing approaches primarily address synthesis of a single component in a system without considering its system-level interactions. Such interactions can be captured by using a more general hardware model (e.g. Petri-net models) to represent the concurrency and communication among components of a system. This approach has the potential to uniformly incorporate both synchronous and asynchronous design styles. However, because of the complexity of analysis techniques, such as checking for reachability or liveness properties [Pet83], synthesis from these models is either lacking or is still in the exploratory stages. Petri-net based synthesis promises to be a fruitful area of further research.

Bibliography

[Aig79] M. Aigner. *Combinatorial Theory*. Springer-Verlag, 1979.

[ASU87] A. V. Aho, R. Sethi, and J. D. Ullman. *Compilers: Principles, Techniques, and Tools*. Addison-Wesley Publishing Company, 1987.

[BBR91] K. Bartlett, G. Borriello, and S. Raju. Timing optimization of multiphase sequential logic. *IEEE Transactions on CAD/ICAS*, Vol. 10(No. 1):pp. 51–62, January 1991.

[BCM+88] R. K. Brayton, R. Camposano, G. De Micheli, R. Otten, and J. van Eijndhoven. The Yorktown Silicon Compiler System. In D. Gajski, editor, *Silicon Compilation*. Addison Wesley, 1988.

[bDG88] Edited by D. Gajski. *Silicon Compilation*. Addison Wesley, 1988.

[BG87] F. Brewer and D. Gajski. Knowledge based control in micro-architectural design. In *Proceedings of the Design Automation Conference*, pages 203–209, June 1987.

[BK87] G. Borriello and R. Katz. Synthesis and optimization of interface transducer logic. In *Proceedings of the International Conference on Computer-Aided Design*, pages 56–60, Santa Clara, CA, November 1987.

[BM89] M. Balakrishnan and P. Marwedel. Integrated scheduling and binding: A synthesis approach for design space exploration. In *Proceedings of the Design Automation Conference*, pages 68–74, Las Vegas, NV, June 1989.

[BN86] J. Burns and A. R. Newton. SPARCS: A new constraint-based IC symbolic layout spacer. In *Proceedings of the Custom Integrated Circuits Conference*, pages 534–539, May 1986.

[Bor88a] G. Borriello. Combining event and data-flow graphs in behavioral synthesis. In *Proceedings of the International Conference on Computer-Aided Design*, Santa Clara, CA, November 1988.

[Bor88b] G. Borriello. A new interface specification methodology and its application to transducer synthesis. UCB/CSD technical report (dissertation), U. C. Berkeley, 1988.

[BRSVW87] R. K. Brayton, R. Rudell, A. Sangiovanni-Vincentelli, and A. R. Wang. MIS: A multiple-level logic optimization system. *IEEE Transactions on CAD/ICAS*, Vol. 6(No. 6):pp. 1062–1081, November 1987.

[CE91] R. Camposano and Wayne Wolf (Ed.). *High-Level VLSI Synthesis.* Kluwer Academic Publishers, June 1991.

[Chu87] Tam-Anh Chu. Synthesis of self-timed VLSI circuits from graph theoretic specifications. Department of EECS (PhD thesis), Massachusetts Institute of Technology, September 1987.

[CK86] R. Camposano and A. Kunzmann. Considering timing constraints in synthesis from behavioral description. In *Proceedings of the International Conference on Computer Design*, pages 6–9, November 1986.

[CR89] R. Camposano and W. Rosenstiel. Synthesizing circuits from behavioral descriptions. *IEEE Transactions on CAD/ICAS*, Vol. 8(No. 2):pp. 171–180, February 1989.

[DDN85] P. Dewilde, E. Deprettere, and R. Nouta. Parallel and pipelined VLSI implementation of signal processing algorithms. In Kung and Whitehouse, editors, *VLSI and Modern Signal Processing*, pages 258–264. Prentice Hall, 1985.

[dG89] A. de Geus. Logic synthesis speeds ASIC designs. *IEEE Spectrum Magazine*, Vol. 26(No. 8):pp. 27–31, August 1989.

[DM92] M. Damiani and G. De Micheli. Synthesis and optimization of synchronous logic circuits from recurrence equations. In *Proceedings of the European Design Automation Conference(to appear)*, March 1992.

[DN89] S. Devadas and A. R. Newton. Algorithms for hardware allocation in data-path synthesis. *IEEE Transactions on CAD/ICAS*, Vol. 8(No. 7):pp. 768–781, July 1989.

[Fre82] S. French. *Sequencing and Scheduling: Introduction to the Mathematics of the Job Shop*. Ellis Horwood Limited, 1982.

[GC91] R. Gupta and C. Coelho. Ethernet controller design. In *private communications*, 1991.

[GDWL92] D. Gajski, N. Dutt, A. Wu, and S. Lin. *High-level Synthesis: Introduction to Chip and System Design*. Kluwer Academic Publishers, 1992.

[GE92] C. Gebotys and M. Elmasry. *Optimal VLSI Architectural Synthesis: Area, Performance and Testability*. Kluwer Academic Publishers, 1992.

[GJ79] M. Garey and D. Johnson. *Computers and Intractability*. W. Freeman and Company, 1979.

[GK84] E. Girczyc and J. Knight. An ADA to standard cell hardware compiler based on grammars and scheduling. In *Proceedings of the International Conference on Computer Design*, pages 726–731, October 1984.

[GM90] R. Gupta and G. De Micheli. Partitioning of functional models of synchronous digital systems. In *Proceedings of the International Conference on Computer-Aided Design*, pages 216–219, November 1990.

[Gol80] M. C. Golumbic. *Algorithmic Graph Theory and Perfect Graphs*. Academic Press, 1980.

[GRVD87] G. Goosens, Jan Rabaey, Joos Vanderwalle, and Hugo DeMan. An efficient microcode compiler for custom DSP-processors. In *Proceedings of the International Conference on Computer-Aided Design*, pages 24–27, Santa Clara, November 1987.

[HB86] D. Hearn and M. Pauline Baker. *Computer Graphics*. Prentice-Hall, 1986.

[HJL+89] J. Huisken, H. Janssen, P. Lippens, O. McArdle, R. Segers, P. Zegers, A. Delaruelle, and J. van Meerbergen. Efficient design of systems on silicon with PYRAMID. In *Logic and Architectural Synthesis for Silicon Compilers*. North Holland, Amsterdam, 1989.

[HPG88] S. Hayati, A. Parker, and J. Granacki. Representation of control and timing behavior with applications to interface synthesis. In *Proceedings of the International Conference on Computer Design*, pages 382–387, October 1988.

[Int87] Intel. *Microcommunications Handbook*. Intel Corporation, 1987.

[Kas89] D. B. Kasle. High resolution decoding techniques and single-chip decoders for multi-anode microchannel arrays. *Proceedings of the International Society of Optical Engineering*, Vol. 1158:pp. 311–318, August 1989.

[KFM91] D. C. Ku, D. Filo, and G. De Micheli. Control optimization based on resynchronization of operations. In *Proceedings of the Design Automation Conference*, June 1991.

[KM90a] D. C. Ku and G. De Micheli. HardwareC - a language for hardware design (version 2.0). CSL technical report, Stanford, April 1990.

[KM90b] D. C. Ku and G. De Micheli. High-level synthesis and optimization strategies in Hercules and Hebe. In *Proceedings of the European ASIC Conference*, pages 124–129, Paris, France, May 1990.

[KM90c] D. C. Ku and G. De Micheli. Relative scheduling under timing constraints. In *Proceedings of the Design Automation Conference*, pages 59–64, June 1990.

[KM91a] D. C. Ku and G. De Micheli. Constrained conflict resolution and resource sharing in Hebe. *INTEGRATION, the VLSI Journal*, Vol. 12:pp. 131–165, December 1991.

[KM91b] D. C. Ku and G. De Micheli. Optimal synthesis of control logic from behavioral specifications. *INTEGRATION, the VLSI Journal*, Vol. 10(No. 3):pp. 271–298, February 1991.

[KM92] D. Ku and G. De Micheli. Relative scheduling under timing constraints: Algorithms for high-level synthesis of digital circuits. *IEEE Transactions on CAD/ICAS*, April 1992.

[Kna89] D. W. Knapp. Manual rescheduling and incremental repair of register-level data paths. In *Proceedings of the International Conference on Computer-Aided Design*, pages 58–61, Santa Clara, CA, November 1989.

[Kow85] T. Kowalski. *An Artificial Intelligence Approach to VLSI Design.* Kluwer Academic Publishers, 1985.

[KP87] F. J. Kurdahi and A. C. Parker. REAL: A program for REgister ALlocation. In *Proceedings of the Design Automation Conference*, pages 210–215, June 1987.

[LBMG89] M. Ligthart, A. Bechtolsheim, G. De Micheli, and A. El Gamal. Design of a Digital Audio Input Output chip. In *Proceedings of the Custom Integrated Circuits Conference*, pages 15.1.1–15.1.6, May 1989.

[LRS83] C. Leiserson, F. Rose, and J. Saxe. Optimizing synchronous circuitry by retiming. In R. Bryant, editor, *Third Caltech Conference on VLSI*. Computer Science Press, 1983.

[LSI87] LSI. *Databook 1.5-Micron Compacted Array Technology.* LSI Logic Corporation, July 1987.

[LSU89] R. Lipsett, C. Schaefer, and Cary Ussery. *VHSIC: Hardware Description and Design.* Kluwer Academic Publishers, 1989.

[LW83] Y. Liao and C. Wong. An algorithm to compact a VLSI symbolic layout with mixed constraints. *IEEE Transactions on CAD/ICAS*, Vol. 2(No. 2):pp. 62–69, April 1983.

[McF86] M. J. McFarland. Using bottom-up design techniques in the synthesis of digital hardware from abstract behavioral descriptions. In *Proceedings of the Design Automation Conference*, pages 474–480, June 1986.

[McF87] M. J. McFarland. Reevaluating the design space for register-transfer hardware synthesis. In *Proceedings of the International Conference on Computer-Aided Design*, pages 262–265, Santa Clara, CA, November 1987.

[MK88] G. De Micheli and D. C. Ku. HERCULES - a system for high-level synthesis. In *Proceedings of the Design Automation Conference*, pages 483–488, June 1988.

[MKMT90] G. De Micheli, D. C. Ku, F. Mailhot, and T. Truong. The Olympus Synthesis System for digital design. *IEEE Design and Test Magazine*, pages 37–53, October 1990.

[MM88] F. Mailhot and G. De Micheli. Automatic layout and optimization of static CMOS cells. In *Proceedings of the International Conference on Computer Design*, pages 180–185, October 1988.

[MM90] F. Mailhot and G. De Micheli. Technology mapping using boolean operations. In *Proceedings of the European Design Automation Conference*, March 1990.

[MMB89] D. G. Messerchmitt, T. H.-Y. Meng, and R. W. Brodersen. Automatic synthesis of asynchronous circuits from high-level specifications. *IEEE Transactions on CAD/ICAS*, Vol. 8(No. 11):pp. 1185–1205, November 1989.

[MPC90] M. McFarland, A. Parker, and R. Camposano. The high-level synthesis of digital systems. *Proceedings of the IEEE*, Vol. 78(No. 2):pp. 301–318, February 1990.

[MSBSV90] S. Malik, E. Sentovich, R. K. Brayton, and A. Sangiovanni-Vincentelli. Retiming and resynthesis: Optimizing sequential networks with combinational techniques. In *Proceedings of the Hawaii International Conference on System Sciences*, pages 397–406, Hawaii, 1990.

[Nes87] J. Nestor. Specification and synthesis of digital systems with interfaces. Ph.d. dissertation, Carnegie-Mellon University, April 1987.

[NK90] J. Nestor and G. Krishnamoorthy. SALSA: A new approach to scheduling with timing constraints. In *Proceedings of the International Conference on Computer-Aided Design*, pages 262–265, November 1990.

[NT86] J. Nestor and D. Thomas. Behavioral synthesis with interfaces. In *Proceedings of the Design Automation Conference*, pages 112–115, June 1986.

[Pet83] J. L. Peterson. *Petri Net: Theory and Applications.* Addison-Wesley Publisher, February 1983.

[PG87] B. Pangrle and D. Gajski. Slicer: a state synthesizer for intelligent compilation. In *Proceedings of the International Conference on Computer Design*, pages 42–45, October 1987.

[PK89a] P. Paulin and J. Knight. Algorithms for high-level synthesis. In *IEEE Design and Test Magazine*, pages 18–31, December 1989.

[PK89b] P. Paulin and J. Knight. Force-directed scheduling for the behavioral synthesis of ASICs. *IEEE Transactions on CAD/ICAS*, pages 661–679, June 1989.

[PKG86] P. Paulin, J. Knight, and E. Girczyc. HAL: A multi-paradigm approach to automatic data-path synthesis. In *Proceedings of the Design Automation Conference*, pages 263–270, June 1986.

[PLNG90] R. Potasman, J. Lis, A. Nicolau, and D. Gajski. Percolation based synthesis. In *Proceedings of the Design Automation Conference*, June 1990.

[PPM86] A. Parker, J. Pizarro, and M. Mlinar. MAHA: A program for data path synthesis. In *Proceedings of the Design Automation Conference*, pages 461–466, June 1986.

[RM88] V. Rampa and G. De Micheli. Computer-aided synthesis of a Bi-dimensional discrete cosine transform chip. CSL Technical Report CSL-TR-88-363, Stanford, August 1988.

[RM89] V. Rampa and G. De Micheli. The Bi-dimensional DCT chip. In *Proceedings of the International Symposium on Circuits and Systems*, pages 220–225, May 1989.

[Rok90] T. Rokicki. Hardware modeling with timed petri-nets. In *Unpublished manuscript*, March 1990.

[SDP90] G. Saucier, C. Duff, and F. Poirot. State assignment of controllers for optimal area implementation. In *Proceedings of the European Design Automation Conference*, pages 547–551, Glasgow, Scotland, March 1990.

[SLP88] L. F. Sun, J. M. Liaw, and T. M. Parng. Automated synthesis of microprogrammed controllers in digital systems. *Proceedings of the IEE*, Vol. 135(No. 4):pp. 231–240, July 1988.

[Sno78] E. A. Snow. *Automation of module set independent register-transfer level design*. Ph.D. Dissertation, Carnegie Mellon University, April 1978.

[Sys88] Mentor Graphic's Silicon Compiler System. *GDT Database and Language Tools Reference*. SCS, May 1988.

[Tan81] A. S. Tanenbaum. *Computer Networks*. Prentice Hall, 1981.

[TDW+90] D. Thomas, E. Dirkes, R. Walker, J. Rajan, J. Nestor, and R. Blackburn. The System Architect's Workbench. In *Proceedings of the Design Automation Conference*, June 1990.

[TIK+83] D. Thomas, C. Hitchcock III, T. Kowalski, J. Rajan, and R. Walker. Automatic data path synthesis. *IEEE Computer Magazine*, December 1983.

[TLW+90] D. Thomas, E. Lagnese, R. Walker, J. Nestor, J. Rajan, and R. Blackburn. *Algorithmic and Register-Transfer Level: The System Architect's Workbench*. Kluwer Academic Publishers, 1990.

[TM91] D. Thomas and Philip Moorby. *Verilog Hardware Description Language*. Kluwer Academic Publishers, 1991.

[Tri87] H. Trickey. Flamel: A high-level hardware compiler. *IEEE Transactions on CAD/ICAS*, pages 255–269, March 1987.

[TS86] C. J. Tseng and D. Siewiorek. Automated synthesis of data paths in digital systems. *IEEE Transactions on CAD/ICAS*, Vol. CAD-5:pp. 379–395, July 1986.

[WC91] R. A. Walker and Raul Camposano. *A Survey of High-Level Synthesis Systems*. Kluwer Academic Publishers, June 1991.

[Wol90] W. Wolf. Rescheduling for cycle time by reverse engineering. In *Proceedings of the ACM International Workshop on Timing Issues in the Specification and Synthesis of Digital Systems*, Univ. of British Columbia, August 1990.

[Wor89] High-Level Synthesis Workshop. Benchmark suite. In *HLSW*, 1989.

[WT89] R. Walker and D. Thomas. Behavioral transformation for algorithmic level IC design. *IEEE Transactions on CAD/ICAS*, Vol. 8:pp. 1115–1128, October 1989.

[Zim79] G. Zimmermann. The MIMOLA design system: Detailed description of the software system. In *Proceedings of the Design Automation Conference*, pages 56–63, June 1979.

Index